Electromagnetic Compossibility

ELECTRICAL ENGINEERING AND ELECTRONICS

A Series of Reference Books and Textbooks

Editors

Marlin O. Thurston
Department of Electrical
Engineering
The Ohio State University
Columbus, Ohio

William Middendorf
Department of Electrical
and Computer Engineering
University of Cincinnati
Cincinnati, Ohio

1. Rational Fault Analysis, *edited by Richard Saeks and S. R. Liberty*
2. Nonparametric Methods in Communications, *edited by P. Papantoni-Kazakos and Dimitri Kazakos*
3. Interactive Pattern Recognition, *Yi-tzuu Chien*
4. Solid-State Electronics, *Lawrence E. Murr*
5. Electronic, Magnetic, and Thermal Properties of Solid Materials, *Klaus Schröder*
6. Magnetic-Bubble Memory Technology, *Hsu Chang*
7. Transformer and Inductor Design Handbook, *Colonel Wm. T. McLyman*
8. Electromagnetics: Classical and Modern Theory and Applications, *Samuel Seely and Alexander D. Poularikas*
9. One-Dimensional Digital Signal Processing, *Chi-Tsong Chen*
10. Interconnected Dynamical Systems, *Raymond A. DeCarlo and Richard Saeks*
11. Modern Digital Control Systems, *Raymond G. Jacquot*
12. Hybrid Circuit Design and Manufacture, *Roydn D. Jones*
13. Magnetic Core Selection for Transformers and Inductors: A User's Guide to Practice and Specification, *Colonel Wm. T. McLyman*
14. Static and Rotating Electromagnetic Devices, *Richard H. Engelmann*
15. Energy-Efficient Electric Motors: Selection and Application, *John C. Andreas*
16. Electromagnetic Compossibility, *Heinz M. Schlicke*

Other Volumes in Preparation

Books are to be returned on or before
the last date below.

24 FEB 1994

0 6 DEC 1995

3 0 NOV 2000

Electromagnetic Compossibility

Applied Principles of Cost-Effective Control
of Electromagnetic Interference and Hazards

Second Edition, Revised and Expanded

Heinz M. Schlicke
Interference Control Co.
Milwaukee, Wisconsin

MARCEL DEKKER, INC.   New York and Basel

Library of Congress Cataloging in Publication Data

Schlicke, Heinz M.
  Electromagnetic compossibility.

  (Electrical engineering and electronics ; 16)
  Includes bibliographical references and index.
  1. Electromagnetic compatibility. 2. Electromagnetic
interference. I. Title. II. Series.
TK7835.S277 1982   629.8'95   82-2499
ISBN 0-8247-1887-9   AACR2

COPYRIGHT © 1982 by MARCEL DEKKER, INC. ALL RIGHTS RESERVED.

Neither this book nor any part may be reproduced or transmitted in any for
or by any means, electronic or mechanical, including photocopying, micro-
filming, and recording, or by any information storage and retrieval system
without permission in writing from the publisher.

MARCEL DEKKER, INC.
270 Madison Avenue, New York, New York 10016

Current printing (last digit):
10 9 8 7 6 5 4 3 2 1

PRINTED IN THE UNITED STATES OF AMERICA

Contents

Preface	vii
Introduction: The Need for a New Approach	xiii

1. SOURCES: TRANSIENTS AND FIELD CONCENTRATIONS — 1

    1.1. Useful Theory — 3
    1.2. Real Sources — 30
    1.3. Counterproductive EMC Measures — 37
    References — 55

2. TRANSFERS: START WITH DEFINABLE COUPLING — 57

    2.1. Conductive Transfer — 58
    2.2. Radiative Transfer — 75
    References — 86

3. RECEPTORS: AN UNSUSPECTED MULTITUDE — 87

    3.1. Technical Receptors (FATTMESS Categorization) — 88
    3.2. Living Receptors — 100
    References — 105

4. SYSTEM ANALYSIS: AN INDISPENSABLE "MUST" — 107

    4.1. Understanding the Systemic Differences — 108

	4.2. Criticalness and Decriticalization	111
	Reference	115
5.	SYSTEMIC CONTROL: THE KEY TO (COST)-EFFECTIVENESS	117
	5.1. Overcoming Handicaps	117
	5.2. The Impact of "Safety First!"	125
	5.3. EMC Commensurate Systems Design	141
	5.4. Comments on Specific Classes of Civilian Systems	162
	5.5. In Summary	166
	References	167
6.	SIMPLE SUPPRESSION: WHAT TO PUT WHERE	171
	6.1. Suppression at Source	172
	6.2. Suppression of Transfer	172
	6.3. Suppression at Receptor	175
	References	175
7.	ABOUT SHIELDING: THE IMPORTANCE OF SIZE AND STRUCTURE	177
	7.1. Frequency Domain	180
	7.2. Time Domain	191
	7.3. Conclusion	194
	References	194
8.	FILTERING FOR EMC: THROW AWAY YOUR FILTER BOOKS	195
	8.1. The Insidious Problem	196
	8.2. The Solution for Indeterminate Mismatch	197
	References	226
9.	GROUNDING AND WIRING, CONTINUED: YOU MUST PLAN AHEAD	227
	9.1. Equivalent Lines	227
	9.2. More About Voltage Gradients	229
	9.3. Key Pointers for Architects and Contractors	232
	References	233
10.	STANDARDS AND TRUTHS: USE THE CODE AND USE YOUR HEAD	235
	10.1. Basic Classes of Standards	235
	10.2. Limits Inherent in Standards	236
	10.3. The Three Truths	237
	10.4. Fragmentation of Standards	238
	References	238

Contents v

11. COMMENTS ON MEASUREMENTS: MOST-NEEDED CORRECTIONS ONLY 239

   11.1. At the Edge of the State of the Art 240
   References 246

12. PROBLEMS AND SOLUTIONS: LEARNING TO CO-THINK EMC 247

   12.1. On Defining a Raw Problem 248
   12.2. Coping with All-Too-Human Frailties 250
   12.3. Problems, Problems: Improve and Check Your Understanding 252
   12.4. Coming on Stream, Maybe 276
   References 282

Index 283

Preface

For the busy engineer, the key question in deciding to read any technical book is: Do I really need it? In the case of this book, the answer is a definite *yes*, but only if the answer is yes to all of the following three questions. (1) Am I concerned about the costs and consequences of downtimes and hazards imparted by industrial/commercial equipment which I design, use, or sell? (2) Are the problems, at least in most cases, caused by vexing electromagnetic interference? (3) Am I interested primarily in getting results, as opposed to only meeting some standards, however marginal they may be? Let me explain.

    This intentionally terse book addresses one of the most pressing, controversial, and misunderstood areas of electrical engineering: the cost-effective prevention of electromagnetic interference and hazards in automated industrial systems. The very fact that such systems are highly sophisticated, due to microprocessors and computers, makes them in many cases very prone to problems, often quite unforeseen, caused by electrical "noise." The apparent unpredictability and untransparency of such interference makes many an engineer rather uncomfortable. This book is intended to convert this very discomfort into ease--in fact, into fun--provided that you like to use your head.

This is achieved by a highly unorthodox approach that is appropriate for the idiosyncrasies of instrumentation and control systems for which conventional RFI techniques are of limited, even dubious, utility. To indicate that we are confronted with a very different set of problems and solutions, we shall denote it by *electromagnetic compossibility*--the *possibility* (of subsystems living) *together*. This stresses the marked difference from conventional, military-oriented electromagnetic *compatibility*, which is a quite different entity. One does not use the same medicine for coughs as for head aches, just because it is medicine.

Let us be clear from the very beginning that we do not want to rehash "established" EMC rules and standards, because:

1. Many of them are simplistically taken out of their systemic context. This applies even to such supposedly simple things as grounding, shielding, and filtering. Such simple prescriptions are easy to follow, but hard to get results with. We shall have no compunctions about killing some very sick sacred cows still in existence here.
2. We are treating highly interactive nonlinear (feedback) control systems which inherently have operating conditions very different from those of the communications systems for which most RFI standards were developed. Hence, many of these rules and standards are just not applicable.
3. We do not intend to provide simple "cookbook" rules, which replace thinking with routine. Hence, this rather intensive book is not for those engineers and technicians who only feel comfortable with predigested step-by-step prescriptions.
4. We rigorously discard trial-and-error methods, which are primitive, unreliable, and have justifiably given a rather bad reputation to conventional EMC.

Rather, we aim at:

1. Fundamental training in thinking systemically in electromagnetic compossibility such that:
   A. We identify the basic differences, individually and interactively, among communications and of control systems in terms of interference sources, transfers, and receptors. Basically, in control systems, the sources are transient and broadband, the transfer is untransparently spread, and unsuspected receptors are necessarily exposed in noisy spaces. In contrast, in essentially linear communications systems, the sources of most RFI are narrowband, the channels are well defined, and the receivers are known and tuned.
   B. We plan the system and the prevention of interference together, because retrofitting for EMC is very costly.
   C. We eliminate the counterpositive effects of eliminating one problem at the expense of unintentionally creating another problem somewhere else.
   D. We accept the mandatory safety standards, like the National Electrical Code, but analyze and control their profound impact on noise immunity.
   E. We accept and consider carefully, in analysis and control, the eight critical parameters contained in the acronym FATTMESS: $\underline{F}$requency, $\underline{A}$mplitude, $\underline{T}$ime, $\underline{T}$emperature, $\underline{M}$ode, $\underline{E}$nergy, $\underline{Si}$ze and structure, and $\underline{St}$atistics.
2. In singling out the essence of our problem--that of organizing EMC disorder--we stress principles and their limitations instead of precepts. We put industrial EMC on a scientific basis.
3. To ensure better understanding and to facilitate flexibility in trade-offs and cost-effectiveness, we juxtapose, in many instances, various causes, effects, and remedies, including their limits and interactions.
4. To ensure brevity, universality, and timeliness, we use non-dimensionalized presentations and nomographs where feasible. An unusually large number of examples provides a good feel for the order of magnitude involved in a particular situation.

Altogether, a very different, innovative, and adaptive approach is taken, which works well for and is very much liked by engineers who think in terms of wholes rather than isolated parts. We get a handle on the apparent interference "mess" by partition, isolation, and redundancy. We partition into noisy, semiquiet and quiet spaces. Concerning isolation, we emphasize indirect differentiation to approximate ideal isolation feasible by electrooptics, which, however, is often still too limiting, due to price, aging, and restricted linearity and range. And, specifically, we straighten out the unbelievable confusion still surrounding grounding, shielding, and filtering--much to the consternation of so-called EMC experts still relying on outdated concepts.

A pilot text, on which this book was based, was originally developed for a continuing education course sponsored by the IEEE, EAB. The course, "Electromagnetic Compossibility: Applied Principles of Cost-Effectiveness Control of Electromagnetic Interference and Hazards in the Civilian Domain," is still being given either through the IEEE (awarding CEAU credits) or directly, in conjunction with confidential in-house consulting services, with the aim of making system planners ready for independent work as soon as possible. Participants in the course represented a wide variety of companies, such as those engaged in the design and production of automotive equipment, and computers, and those in fusion research, industrial and process control, and electric power distribution systems, all using microprocessors, etc.

The treatment of the admittedly complex and controversial problems covered in the course generally turned out to be appreciated by those for whom it was intended, although more traditional thinkers occasionally disliked it. The book as it is now offered is essentially the result of revision and addition to the first text. When the course material was adapted for this text, some significant changes were made. Chapter 12 was added to facilitate self-study. It has a three-fold purpose: (i) It attempts to further explain and stimulate much-needed systemic (holistic) thinking--in contrast to the routine thinking that presents the

Preface

greatest danger to good EMC. (ii) It provides more than 100 problems and solutions of various degrees of difficulty, to check and augment the reader's understanding of the pervasiveness of the FATMESS criteria. (But you have to apply your own brain!) I am grateful to many course participants who were not aftaid to ask "stupid" questions now included in Chapter 12. And finally, (iii) It marshals and assesses some still embryonic solutions which, in all liklihood, will soon significantly affect the cost-effectiveness of EMC control. For a clearer presentation, the figures and tables were, where necessary, amended and simplified as suggested in a review of the text by A. H. Sullivan, Jr. (See IEEE Spectrum, Oct., 1979, book review, or IEEE EMC Newsletter, Winter, 1980.) A comprehensive index has also been provided. In additic I have corrected typographical errors and some errors of fact that were pointed out by Dr. Prit Chowdhuri, Los Alamos National Laboratory. I thank Dr. Chowdhuri for his constructive criticism.

I am also grateful to Dr. W. Scott Bennett, who pointed out the potential uncertainty of far-field measurements made according to VDE standards when I gave the course for Hewlett-Packard at Fort Collins, Colorado.

Heinz M. Schlicke

Introduction
The Need for a New Approach

This candid and rather unorthodox text, aiming at the edge of the state of the art, is concerned with the presently available facts and principles, often misapplied, which practicing engineers must know if they want to prevent harmful electromagnetic effects upon technical systems and human beings. Here the emphasis is on civilian noncommunications environments, where so far the need for such prevention has been rather limited, in contrast to military environments, where preventive action against radio-frequency interference (RFI) has long been considered indispensible and has been greatly standardized, although not always with a critical mass of brain power. Now, the proliferation of solid-state devices handling power and multiplicity of information in close proximity necessitates judicious preventive measures against harmful electromagnetic interference in the civilian sector. But because of distinctly different operating conditions, the military experience cannot be transferred with impunity to the industrial and commercial world. The failure to recognize the significant differences in operating conditions is the reason for much frustration caused by the failure and/or high price of anti-interference measures based on standards heretofore established primarily for communications systems. The systems we are concerned with here, by their very nature, often violate the principles so well developed for and applied in communications systems (intentional versus incidental "radiation"; definable versus spread transfer; exposed receptors, etc.).

Under ideal conditions, automation of processes of all kinds provides economy and eliminates human error and hazards. This idealization applies to industrial control, automatic rapid-transit systems, and automated patient care, to name just a few classes of examples. Yet, in reality, the more the automation is refined [e.g., by employing LSI (large-scale integrated circuits)], the more errors, resulting in costly downtime or hazards, may be introduced by electromagnetic interference (EMI), also called, rather restrictively, electrical noise. Natural and artificial phenomena are encountered. Computers, minicomputers, and microprocessors and programmable logic controllers on a chip are essential, but very interference-prone entities are involved. Signals coming from sensitive sensors may be falsified by noise from within or from outside a particular system. The most critical of such noises is often a switching transient. But natural causes such as lightning or static electricity can also pose problems. Natural receptors of EMI are human beings, who are sensitive (beneficially or adversely) throughout the whole frequency spectrum. This includes some very unsuspected "side effects."

Thus makers and users of electrical equipment are becoming more and more concerned with electromagnetic compatibility (EMC), that is, planned prevention and control of electrical noise. System designers, being nonspecialists in EMC, need assistance to make the right decisions and compromises in the multifaceted and multiple-choice aspects of EMC. Fiber-optics technology, if properly applied, offers significant cost-effectiveness of EMC, but it is no cure-all.

In many quarters, EMC is considered with a certain amount of disdain. There is some justification for the attitude that at present EMC is a somewhat secretive empirical art and is operating

Introduction

with committee-generated standards and specifications of often doubtful utility. We discuss this in detail throughout the text. But even assuming that the standards are correct and applicable, at least to a degree, disappointment with conventional EMC is caused primarily by unjustifiable expectations. As long as the systems engineer looks at EMC as a collection of routine guidelines (cookbook approach) which try to subsume multipurpose, overlapping aspects under one common denominator, and which the engineer can apply without much thinking, he or she will not succeed, except, perhaps, at a high cost. Moreover, the EMC-wise control systems are much less explicit than communications systems, as closed-loop control systems are quite different from essentially open-loop systems. The statements just made require elaboration.

To differentiate clearly between conventional, formalistic, military-based electromagnetic compatibility and our more flexible approach, based on synergistic coapplication of theoretical principles, the latter shall be denoted as electromagnetic compossibility (the latter word, according to Webster, meaning the possibility of living together). This is not a play on words; rather, a basic difference of philosophy is involved. Electromagnetic compatibility, as used in this context, implies rigidity, formalization, bureaucracy; whereas electromagnetic compossibility stresses flexibility and systemic thinking. Very systematically we shall develop comprehensive, noncontradictory ways of EMC thinking. But still, electromagnetic compossibility necessitates an intuitive attitude if it is to solve economically the growing problem involving the coexistence of proliferating systems without detrimental interaction.

Certainly, there are a myriad of government specifications, standards, regulations, and the like, pertaining predominantly to "intentional" narrow-band "radiation." But what good are such prescriptions in civilian environments, where we are confronted with an intrinsically different situation: many broadband "incidental sources" coupling compoundedly into many highly susceptible receptors?

There is another fundamental difference between governmental and nongovernmental EMC activity: documentation of EMC work done. EMC engineers working for the government or under government contract have to meet many standards, specifications, procedures, and so on, and have to document their work elaborately. The more they do so, the more respectable or profitable is their work and the less they and the contracting officers can be held responsible if later something should go wrong. At least, engineers think so.

In private industry, in contrast, the key criterion for good EMC work is how to make the system immune to interference and hazards with minimal overall costs, short and long range. One has to weigh the costs, for example, of downtime on production lines (let us say $2000 per minute) against the cost of suppressing electrical impulses that can cause such downtime. Naturally, this emphasis on costs does not imply advocacy of violating existing regulations concerned with hazard prevention (e.g., National Electrical Code or Underwriters Laboratories' Standards).

For lack of relevant (civilian) EMC information, then, the IEEE Industry Application Society started working on a "Noise Guide" (called "Guide for the Installation of Electrical Equipment to Minimize Electrical Noise Inputs to Controllers from External Sources"). The "Noise Guide" was developed under the able leadership of G. Younkin of Giddings and Lewis, Fond du Lac, Wisconsin. The "Noise Guide" purposely confines itself to the aspects of installations and does not address the broader problem of designing "any" system from the viewpoint of electromagnetic compossibility, which is the key concern of this text. My work with the "Noise Guide" Committee, still continuing, prompted me to write this text, which is broader in scope than and complementary to the "Noise Guide" and related documents which we shall refer to in appropriate places.

The objective of this text is to really understand the underlying principles and their criteria so that you can make your systems coexist. In contrast, the objective of the "Noise Guide" is: "When installing a control system, follow my prescriptions, and you

Introduction xvii

will have no problems" (thereby tacitly assuming that the system per se has already been designed for EMC--inter- and intra-system-wise).

Other strong indicators for the need for a frank text on non-military EMC come also from my experiences as course director (with R. Ficchi as codirector) on EMC courses for the Center for Professional Advancement and from many curious responses to the rather cursory article "Getting Noise Immunity in Industrial Control" (IEEE Spectrum, June, 1973; co-author, Odo Struger).

A similar talk given in Moscow in April 1974 before the Popov Society also found a very responsive and information-hungry audience.

Feedback from practicing engineers, obtained during EMC courses and consulting sessions held for industrial concerns, clearly points to the need for a preferential treatment of conceptual difficulties facing and often fazing the engineer not initiated in electromagnetic compossibility.

Product liability is another growing major concern. Until recently, only obvious negligence or deceit could be held against a manufacturer. In general, the rule "let the buyer beware" was dominant. But the consumer movement has converted this into "let the manufacturer be sued." Manufacturers are now liable under the strict tort liability concept. They should not leave the consequential design for safety to rigid safety engineers who go "strictly by the book." There is no doubt that guidelines, standards, and codes carry much weight with juries as a measure of the preventive care exercised by manufacturers. Such codes (e.g., the NEC) are periodically updated and at any given time, represent "what we know best, as of now," as a general rule for a categorized set of related circumstances. In specific cases, however, there are often particular concomitant conditions that may override what was initially considered the only or dominant "universal" statement of concern. Only by looking at the whole of a particular situation, from different aspects, and from the views of several unbiased, alert experts can we come close to an appropriate assessment of safety measures, because safety is not threatened only by such things as lightning or ground faults but more and more by catastrophic systems malfunction caused by EMI. Rigid adherence to imperfect standards is sometimes the worst one can do, as we shall demonstrate on a number of occasions.

After defining the existing and growing needs, let us delineate what this short book on EMC is about and what it is not about. We want to marshal, in handy but not simplistic form, the principles and techniques for cost-effective prevention of electrical interference and hazards as encountered primarily in industrial and commercial systems. EMC for military and communication purposes is quite a different affair and has already been treated in many texts. Nevertheless, the engineer concerned with military EMC should also benefit considerably from the text.

Although there are always some system-specific requirements for each class of systems, there is great commonality of the basic causes of, and remedies for, electrical interference and hazards, as will soon become obvious. This very fact facilitates the tentative establishment of a theory of EM compossibility in spite of its interdisciplinary and practical nature. Theory, in this context, turns out to be a very practical thing in that it permits us to work predictively from principles. We shall borrow from textbooks on Fourier transform, circuit theory, electromagnetic waves, information theory, and other subjects. Thus we shall not bother to rederive, like regurgitating cows, those equations and relations that are in many other places. However, we shall take many a generally accepted equation with a great lump of salt (not only a grain), or derive new relations where necessary. Also, since the text is written for practicing engineers who finished formal schooling some time ago, we avoid complex mathematics.

The assumption upon which a particular equation was derived originally may not be valid under the circumstances given, necessitating a delineation of the range of validity, often resulting in a drastic revision of established usage (e.g., large frequency bands, bias and amplitude differences, severe mismatch, etc., are scarcely

Introduction   xix

encountered in system design). Systems designers are apt to view such conditions as uncritical. Much to their chagrin, they will learn otherwise when interference strikes their system, which turns out to be a very faulty communications system.

A system made inoperative or unsafe, by EMI or otherwise, is a dangerous or at best a useless thing. In good EMC practice, then, one does, with a carefully weighted minimum of overall costs, all that is necessary, but not more (as one does not provide a car with more than one spare tire).

But more comments relating to cost-effectiveness seem to be in order. Although on occasion we make some rough estimates of comparative costs, we apply the term cost-effectiveness more in an implicit sense than providing specific data for cost accountants. By "implicit" is meant that overall cost minimization can be obtained from an analysis of critical EMI relations and circumspection regarding a prior planning of options of EMC remedies such that they become an integral part of the system, preferably so much so that the cost of the system is reduced. This can be done by proper partitioning and judicious use of isolation (see Chapter 5), particularly in respect to exposed receptors.

Beyond such different, highly beneficial systems planning, we must also change our thinking about "standard," "simple" EMC remedies such as filtering, shielding, and grounding. We shall see that we are confronted with problems which we may not think are problems--rather "old hat," treated and solved in textbooks for 30 or 40 years now. Well, we shall see!

1. The power feed lines may carry switching transients through the entire system they feed. Filters can prevent this if and only if they are properly designed for the unavoidable and severe (often indeterminate and inconstant) mismatch conditions encountered. Power feed lines, their sources and loads, are designed for the sole purpose of high efficiency at 60 Hz. They are not transmission lines with constant characteristic impedance and matched generator and load impedances over the whole frequency

range from dc to microwaves, as textbooks on filters assume.
The possible resulting dysfunction of such filters may have
very undesirable consequences, which may occur although the
filter meets rigid but unrealistic specifications. The theory
of filters working into indeterminate interfaces, as developed
in Chapter 8, forms the basis of worldwide modification of
filter standards (CISPR, NATO, Military Standard 220A, etc.)
which will become effective about 1980.

2. Nonlinearity, for instance the saturation of magnetic or dielectric materials under large current or voltage bias or field concentrations, may have wholly unexpected effects. Their prevention (or exploitation) will be discussed throughout the book whenever such measures become necessary.

3. Shielding is another case in point. Calculated shielding effectiveness is often invalidated by leakage through seams, doors, or holes which are required for practical utility. We look at shielding from an unconventional point of view, emphasizing the significant role of the size of the shielding enclosure. This in turn affects cost considerations. The reduction of rise times of pulses by shields requires careful analysis. Absorption by lossy materials is often important and economical.

4. Misapplication caused by misunderstanding and/or deterioration (e.g., caused by corrosion) of "ground" and "common" systems are a matter of great concern. Such systems do not always provide the hoped-for protection to reduce shock, reduce excessive fault currents, or eliminate control errors. Again, officially prepared and sanctioned "standards" or codes will be found not always to be reasonable or sufficient recommendation.

These examples, typical for what is involved in practical EMC work, are also indicative of another distinctive characteristic of EMC: the mushy, indeterminate nature of input data and operating conditions. This makes exactly calculated output data look like pretense. Still, we must be able to quantify our statements, although we lack, in many cases, a meaningful data base. It makes

Introduction   *xxi*

no sense but is often done for "effect," to make EMC calculations exact to the fourth decimal place. In fact, this is as phony as high-accuracy sales or budget forecasts, for which the high accuracy of the computer is supposed to assure high accuracy of highly uncertain assumptions. Rather, the exacting work must be in defining the underlying problem and its boundaries in such a manner that its solution is economically manageable, and not treating the wrong problem according to inapplicable specifications or models. Realistic worst-case conditions, approximating the real world, not for textbook convenience, must be delineated. In particular, the use of statistical mean values can be as dead wrong as it is wrong to assume that all people of a nation will die at age X because it is their average life expectancy.

We must decide where exact calculations are appropriate and where the determination of the order of magnitude (OOM) is reasonable. Many practical examples are interspersed in the text so that readers can slowly build up their ability to see what counts and which simplifying model is appropriate to arrive at realistic estimates. In fact, the abundance of numerical examples, in particular of comparative ones, renders the text highly suitable for self-study.

Since many contrary objectives have to be met simultaneously, it often happens that a specific anti-interference or anti-hazard measure, while eliminating the problem on hand, causes another interference problem, even more serious than the one disposed of. We devote quite some time to such contrapositive effects inasmuch as they are of significant didactical value. In this context we cite examples severely affected by nontechnical aspects.

In our effort to extract the essential, we are rather selective in our presentation. Very much contrary to a conventional textbook, the treatment of the various topics is purposively quite unequal. Engineering concepts and methods that are well known, noncontroversial, and well documented elsewhere are treated very briefly, if at all. They are more or less given as broad-brush hints for perspective and continuity only. (We have no intention to reproduce the excellent application notes of manufacturers of such solid-state devices as instrumentation amplifiers, isolators, and V/F converters, as they are readily available.) In contrast, emphasis (only seemingly disproportionate) is put on these topics for which conventional approaches do not work satisfactorily. This applies to the many cases where, because of the seeming complexity of the situation, simplification is required for tractability, but the wrong simplification (model) is used.

For perspective and facilitation, many a table places side by side various aspects of generic topics in picture-book form, combining diagrams, sketches, and formulas. Whenever feasible, we use normalized, dimensionless scales, coordinates, and parameters; define the range of validity; and point out asymptotic approximations. Nomographs are included for practicality. The MKS(r) system replaces the medieval inch-pound system.

After these orienting remarks about the need existing, the objectives tried for, the idiosyncrasies specific to EM compossibility, and the general approach dictated by the desire for real utility of the text, the plan of the book, as shown in the accompanying illustration, does not need much explanation.

As the chart shows, the text consists of three major groups: analysis, control, and measurements. The first group (Chapters 1 to 4) deals with EMI analysis. We must know why things happen to take preventive or corrective measures. Chapters 1-3, respectively, represent a "data bank" on the three key elements comprising the EMI link: sources (with strong emphasis on impulsive noise), transfers, and receptors, including human beings. This refers to EMI in general, both from a mathematical and from an empirical point of view. "Data bank" may be an exaggerating misnomer for something unfortunately still incomplete. A better description might be "pertinent data available so far." But the data bank provides a "handle" for calculations or estimates; equally important, it demonstrates the diversity of often unconceived EMI problems.

Introduction

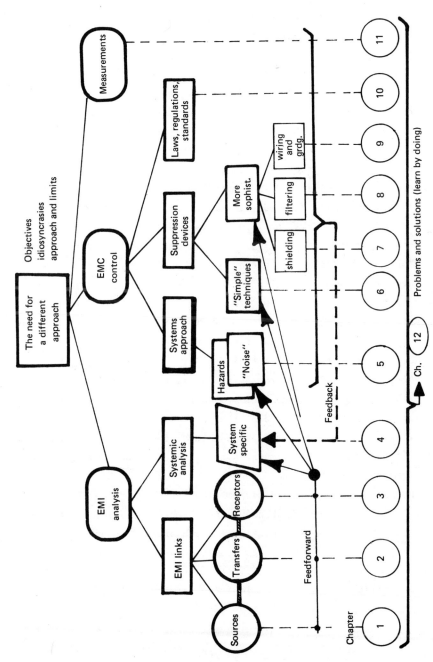

Plan of the Book

In contrast to the specifities treated in Chapters 1 to 3, Chapter 4 is a systematic attempt to analyze the control system as a whole. This is best done by rearranging the diagrams and flow graphs of the systems designer in terms of EMI/EM compossibility. This rethinking of the operating system as an EMI system is the key to cost-effective EM compossibility and must include continual, vicarious feedback between analysis and control options.

For this purpose we must acquire a clear understanding of the fundamental differences that noise plays in a readily definable communications system and in a complex, widespread control system. In the latter, many hundreds of sensor and control links, on the one side, may be hooked up to a corresponding number of controlled functions (which are widely distributed and often generate noise), and on the other side, to a common data processing unit--or to more or less autonomous satellite subsystems (microprocessors) remote from the central unit. Thus broadband, intensive noise sources and sensitive noise receptors are tightly embedded. Here the "spread" transfer of interference is of particular concern because of its opacity. We shall untangle this mess by a systematic approach involving a methodical, progressive elimination of criticalness.

The second major group belongs to control (Chapters 5 to 10), already very much initiated by the preceding chapters. Thus the cardinal chapter, Chapter 5, is a natural continuation of Chapter 4. In fact, there is so much internal feedback between analysis and control that they are often barely separable. Hazard prevention and EMC-commensurable system planning must be coordinated. Here we stress the comprehensive and unambiguous elimination of contrary objectives and conditions which are very confusing to the uninitiated and which are caused by the large ranges of frequencies, amplitudes, and dimensions encountered in large systems and which are aggravated by antinomies imparted by mandatory safety codes.

Chapter 6 treats relatively simple but effective suppression means. Chapters 7 to 9 deal with shielding, filtering, and wiring and grounding, respectively. They continue and/or conclude discussions initiated earlier. They stress realistic new solutions

Introduction xxv

and do not rehash "time-honored" but unjustifiable oversimplification based on convenient but not EMC-commensurate boundary conditions. As already mentioned, these three means are supposedly simple EMC techniques about which misconceptions are widespread and about which some misleading standards are still in effect. And even worse, such erroneous information is still promulgated in current EMC texts and courses.

There is no cookbook approach to EMC; neither is there much sense in trial-and-error methods or, on the other extreme, in highly formalized prediction methods. A good EMC engineer cannot afford, in the industrial environment, the formalistic matrix approach that military EMC engineers apply justifiably in their essentially "intentional" environment. And even after the system is carefully designed for EM compossibility, its installation must be supervised with equal care lest one run the risk of electrical noise and hazard problems caused by inappropriate wiring or grounding in the field. Yet with proper isolation of all critical receptors (see our discussion of an ideal system in Chapter 5), such on-site control is greatly facilitated.

Unless already discussed in previous chapters, laws, standards, specifications, and so on, whether official documents of self-imposed regulations, are very briefly dealt with in Chapter 10. We certainly have no intention to reprint, for example, the many applicable interference and safety regulations but tacitly assume that the reader is familiar with them. Engineers must know codes and meet them for legal and commercial reasons. But, as is the case with all generalizations in compounded situations, they may not fit or may exclude some pertinent facts of the case, as demonstrated on several occasions in the book. Readers must always bear in mind that they are not working on a singular-purpose, well-structured concern, but must meet many coexisting needs, while laboring under constraints, obstacles, and uncertainties. Routine prescriptions made for limited purposes do not solve unstructured problems.

Brevity also holds, by necessity, for Chapter 11, measurements (constituting group III). Pertaining to measurements, much information useful in terms of the objectives of the book is still rather embryonic or in flux. As we cannot wait until these things are settled, we provide a brief description, including possible sources of mistakes, for selected cases where conventional measuring or testing methods are not applicable.

Thus, while this short book is not about design systems per se-- it is even not about military-oriented EMC--it is intended for the engineers who (a) want to make industrial and commercial systems immune to interference and hazards with a minimum of cost and learning errors and who (b) --if a system fails due to EMI--cannot blame imperfect standards and codes however diligently implemented. They will be the people who get "hung."

Our introduction is perhaps most relevantly concluded with the comments made by the IEEE, EAB (Educational Activities Board) Evaluation Committee on the course material submitted when it was about half-completed: "It is [our] opinion that this is one of the more difficult and one of the least understood areas of design and manufacture. Any scientifically-based work in the area is welcome. The field is absolutely filled with misleading and erroneous data, procedures, and specifications. Dr. Schlicke seems to have the right approach to the material; and [we] believe that if the lecturer (or lecturers) is of the philosophy contained herein, the short course would be provocative and controversial--two things which the subject badly needs."

Heinz M. Schlicke

Electromagnetic Compossibility

1 Sources
Transients and Field Concentrations

When microprocessors were first introduced into school buses (to operate the antiskid devices) without the benefit of electromagnetic compatibility (EMC) planning, severe hazards were invited: A radio transmitter of a police car or radio amateur close by would actuate the microprocessor such that the compressed air was soon exhausted and the brakes of the school bus became inoperative. Thus a supposedly protective device was actually a potential killer. Now, microprocessors are carefully shielded and filtered. This was not done without much-needed consulting since design engineers (not being EMC engineers) installed capacitors such that they acted as inductors and provided shields that were not much of a shield at the frequencies encountered.

Such radio frequency interference (RFI) caused by intentional radiation is easily predictable and should have been predicted. And in this case, retrofitting was a rather simple affair. But interference is less obvious to analyze and is harder to control if switching is involved.

Switching of electromagnetic energy, be it intentional as for control purposes, or incidental as in the case of lightning or faults, is a dominant source of interference and hazards. The most disturbing (to digital equipment) interference is caused by the transitory part of the switching (dynamic), whereas most hazards are dangerous because of the steady state condition being switched on (static). At any rate, whether we can influence the switching or not, we must have a clear understanding of what happens so that we can handle the problems arising.

In the case of *interference sources*, we are interested in how amplitudes and d/dt's are affected by circuit elements located in the immediate vicinity of the switch (one-port approach; how interference is changed in transfer through a two-port is discussed in Chapter 2).

In the case of *hazard sources*, field concentrations are of major concern. (Again, the coupling of hazardous amplitudes is deferred to Chapter 2, although it is often difficult to divorce sources from transfers.)

Our classificatory system set up to handle the complex and manyfold sources of noise and hazards is pragmatic: (1) we shall extract, from the abundant literature on the Laplace transform, the essence of the theory of transients. Surprisingly simple, but very handy models for field distortions are considered in Section 1.1. (2) We shall describe the gist of real sources of switching interference and of hazards ranging from minute random energy fluctuations (bandlimited white noise) limiting the amplification of signals to the enormous energy bursts let go in electromagnetic pulsing (EMP) or lightning. To get the proper perspective, material-caused sources are included in Section 1.2. (3) We shall marshal representative examples of counterpositive EMC measures. They are valuable in that they teach the need for circumspect, systemic (looking at the whole system in terms of long-time events) planning instead of addressing a specific problem or standard in isolated form without considering equally important circumstances such that new problems may arise, sooner or later, from convenient but myopic reliance on standard procedures (Sec. 1.3).

1.1 USEFUL THEORY

1.1.1 Linear Switching Theory

Before we discuss our main concern, the role of time constants in the generation of transient interference, we shall run over some very important, related facts which we are supposed to know. They should be ingrained in our minds as automatically as the alphabet (if not, the reader should pick up a book on the Laplace transform, such as [1]).

1. A *discrete frequency spectrum* (harmonics, if a single causative frequency is involved; or combination frequencies, if several frequencies are involved) is caused by a nonlinear element and continuous sinusoidal excitation (Fourier series). (Consider switching power supplies: repetitive pulses do not contain frequencies below the switching frequency.)
2. A *continuous frequency spectrum* (amplitude reciprocal to frequency for step function; constant amplitude for all frequencies for a Dirac pulse--otherwise bandlimited for pulses of "nonzero" width) is generated by discrete (single) switching events (Fourier transform).
3. A (single) time function, having an upper spectral limit at frequency f_u, *does not have to be continuous*, but rather is completely defined if *sampled at intervals*, $T_s = (1/2\ f_u)$ seconds. The large time spaces between sampling points are free for other operations (sampling theorem).

It will also help our understanding of switching transients if we delineate the circuital conditions for which an ideal switch (no arcing, bouncing, etc.) will *not* create switching transients (delay, overshoot, etc.), but is an ideal on-off step function. That is the case if, and only if, the switch switches an ideal resistor: namely, that rare entity that is independent of frequency, amplitude, time, and temperature. (Even an incandescent lamp is a far cry from being ideal even at low frequencies: In the first few milliseconds after being switched on, its resistance increases to about 10 times its cold value; thus the high inrush current settles relatively fast into the steady state current.) But the main reason why an ideal switch does not switch ideally is the unavoidable presence of storage elements in the immediate neighborhood of the switch. Each piece of wire we use has an L and a C. We shall now see how these reactive elements affect the switching.

1.1.1.1 Step Functions and Time Constants

Practically Independent Time Constants

There is always a delay to charge or discharge an energy storing element, as shown in Figure 1.1, for switching on a storage element over a resistor. For the inductor, the current rises with $(1 - e^{-t/\tau_\ell})$; τ_ℓ = L/R. For the capacitor, the voltage rises with $(1 - e^{-t/\tau_c})$; τ_c = CR. For switching on, then, a voltage in a series reactance/resistance circuit:

$$V_c(0_+) = 0 \quad \text{and} \quad I_\ell(0_+) = 0$$

where 0_+ is the time immediately following the closing of the switch. Now, that means:

$X_c(0_+) = 0$ You cannot drive a step voltage into a C.

$X_\ell(0_+) = \infty$ You cannot drive a current step into an L.

In contrast, the inductor voltage or the capacitor current jump instantly to their maximum value and decay with $e^{-t/\tau}$. This initially infinitely steep jump seems, at first sight at least, physically impossible, but it only seems so. No energy is involved at that moment; $(CV^2/2)$ or $(LI^2/2)$, respectively, is still zero.

Example 1.1 If a resistor, let us say of purely 100 ohm (assuming that such a thing can exist), is switched on a battery of 1 V, the response is immediate: The current will be 10 mA, the voltage 1 V. If, however, an inductor of 10 µH is put in series, the current will rise with $\tau_\ell = (10^{-5}/10^2)$s = 0.1 µs; whereas the voltage across L will jump immediately to 1 V and decay with the same time constant, to be, e.g., 3 Np (Np = neper = e^1 = 8.69 dB) below 1 V at the time

Useful Theory 5

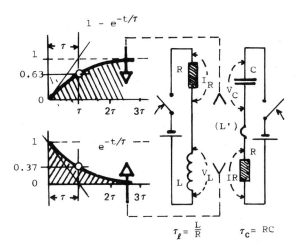

Figure 1.1 Single Time Constants

$3\tau_\ell = 0.3$ μs. But in reality, the inductor voltage will not rise instantly to full value. For initiation of the charging process, we can disregard the L momentarily; instead, we have to take into account an unavoidable distributed capacitance (across R) of say 10 pF. Thus the theoretically infinitely fast rise will actually be finite:

$$\tau_c = 10^{-11} \times 10^2 \text{ s} = 1 \text{ ns}$$

It is a great convenience that we can treat the two time constants independently (even though they refer to the same R). We must, however, realize that separate treatment of the τ's is permissible only if the two time constants are very different (ratio larger than 100:1, to obtain insignificant error) and if the Q is much smaller than 1 (see "Joint Time Constants" below).

Example 1.2 The conditions just stipulated hold also for this important example: the slowing of a steep wave front for better protectability. An open-circuited trafo with a transfer inductance of 0.26 H is fed by a 60 Hz line of $Z_0 = 500$ ohm. A very steep wave front, approximated by $(10 \text{ kV}) \cdot e^{-t/1 \text{ ms}}$ (which means slowly decaying),

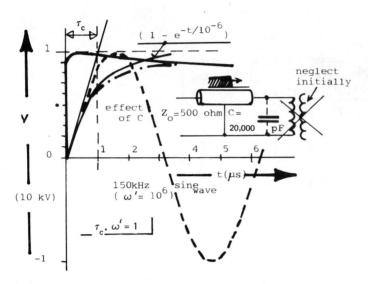

Figure 1.2 Low-Pass Filter Effect Flattens Steep Wave Front (from "Wirbelstroeme und Schirmung in der Nachrichtentechnik," H. Kaden, Technische Physik in Einzeldarstellurgen, Bd. 10-2, völlig umgearb. Aufl., 1959).

is moving along the line toward the trafo. This is equivalent to switching on 10 kV, which would cause the trafo to brake down if its protective device did not respond fast enough. Luckily (Fig. 1.2), considering only the initial impact of the wave, we can neglect the trafo [$X_\ell(0_+) = \infty$] and take only its distributed capacitance (say 2000 pF) into account. That means that the dangerous, very steep wave front will be "flattened" by the time constant $CR = 2 \times 10^{-9} \times 500$ s = 1 μs. This corresponds to the steepest rise [$d(\sin \omega_0 t)/dt = \omega_0 \cos \omega_0 t$] of $\omega_0 = 10^6$ or 150 kHz. The distributed C, then, together with the Z_0 of the line, acts like a delaying low-pass filter with a cutoff frequency of 150 kHz.

Joint Time Constants

If there is only one L and one C (and we shall confine ourselves to this single-frequency condition because it shows the essence of the phenomena encountered), oscillations at a pronounced resonance frequency are possible. Two broad categories of causes exist:

Useful Theory

1. *Negative resistance or positive feedback of active devices (continuous or transient).* A negative resistance may be a falling V/I characteristic, as is the case in an electric arc, or equivalently the negative R of a regulator caused by feedback. Unless the negative R is compensated by losses, +R, in the system, the circuit will oscillate, with amplitudes increasing. The increase is stopped only by nonlinearities arising (unless destroyed by breakdown or heat before a steady state is reached).
2. *Switching of high-Q passive circuits (transient).* For passive circuits we must distinguish between three classes of worst-case oscillatory transients (Table 1.1).
    A. *Insertion gain.* The source contains a frequency (often a harmonic) at which the circuit resonates. When switching on the source, the voltage (or current) of the resonance frequency will increase to Q times its source value. Typical for such conditions are eigenresonances of filters (Chapter 8). For instance, if a 60 Hz source contains a 10 V harmonic and the Q of the filter is 35 (not an uncommon value), the amplitude of the harmonic will increase to 350 V.
    B. *Ringing.* Ringing is typical of a *normal transient* (circuit contains no stored energy). If a step voltage is applied, the voltage (e.g., of the capacitor for a series resonant circuit) will rise to nearly twice the source voltage, with a maximal rise d/dt of $2V_i \omega_0$. We shall discuss normal transients in more detail shortly.
    C. *Abnormal transient.* An abnormal transient is caused by switching energy stored in an L or C, typically switching off current-carrying inductors or charged capacitor banks. Case c of Table 1.1 illustrates the case of a switched-off inductor: The magnetic energy $LI^2/2$ still in the L at switching time can then discharge only into the distributed C of the inductor, until the C is fully charged to $CV^2/2$ (here V is the peak voltage). The energy exchange between

Table 1.1 Oscillatory Transients

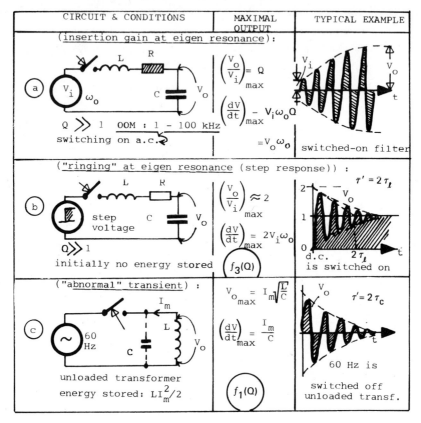

L and C continues with diminishing amplitude (for a series resonant circuit with $e^{-t/2\tau_\ell}$, for a parallel resonant circuit with $e^{-t/2\tau_c}$) until all energy is dissipated in losses. As $LI_m^2/2 = CV_0^2/2$, $V_0 = I_m Z_0 = I_m (L/C)^{0.5}$ and $(dV/dt)_{max} = \omega V_0 = I_m/C$.

Example 1.3 Switching off an (open) trafo with L = 10 H, C = 1000 pF, I_m = 1 A results in V = 1 A $(10/10^{-9})^{0.5}$ = 100 kV (peak) and $(dV/dt)_{max}$ = 1 A/10^{-9} F = 1 V/ns. This example shows that abnormal transients may generate excessive voltages with steep d/dt's such that arcing (and chopping; see later) across the opening contact occur, unless the C is very much increased (see arc suppression, later!).

Now, let us go back to normal transients as they affect step functions in the presence of a dominant resonance frequency, be it series or parallel. Table 1.2 and Figure 1.3 condense graphically what are otherwise not immediately transparent Laplace transforms. Table 1.2 shows the three most useful basic relationships--$f_1(Q)$, $f_2(Q)$, and $f_3(Q)$--with Q as parameter, and sufficient examples for the conditions for which they apply. A little reasoning, keeping in mind that series and parallel circuits are dual, is sufficient for the application. For example, in $f_3(Q)$ we recognize the ringing, with maximally double amplitude as discussed in Table 1.1 for high Q. The only thing we have to know is the Q of the circuit. (Watch out! There are reciprocal Q terms for series and parallel resonant circuits.) $Q = 1/2$ is the critical damping. For $Q < 1/2$, $f_1(Q)$ and $f_3(Q)$ are nonoscillatory (overcritical damping). In applying these very convenient graphs, we have to be careful that the circuit is really in the proper dual form sketched in Figure 1.3, lest we make a serious mistake.

If the actual circuit is not one of the dual ones shown, the following conversion must be made. If there is R_p parallel to a reactance in a series resonance circuit, or a resistor R_r in series with a reactance of a parallel resonance circuit, we can convert to the proper circuit by using the relation $R_p R_s = X_0^2$. For large Q, the conversion error is insignificant.

To demonstrate the broad applicability of the generalized $f(Q)$ graphs, let us stay with $f_3(Q)$ a moment longer. The excessive ringing of LC filters used for SCR rectifier bridges (see Sec. 1.3) is an inverted $f_3(Q)$ function (turnoff). Recovery voltages of fault-clearing circuit breakers are also based on $f_3(Q)$ functions.

The equations displayed in Table 1.2 point out the interesting relationships of joint time constants (both τ_ℓ and τ_c refer to the same R; the R is "owned" by both time constants). A rather striking fact is obvious: If we wish, we can make all our calculations

Table 1.2 Joint Time Constants (from Ref. [1])

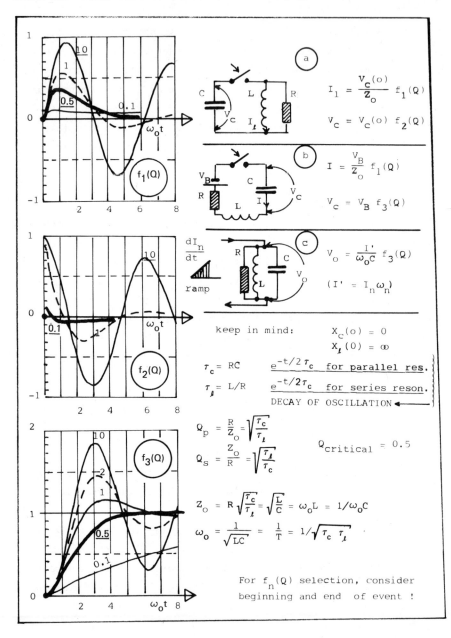

Useful Theory

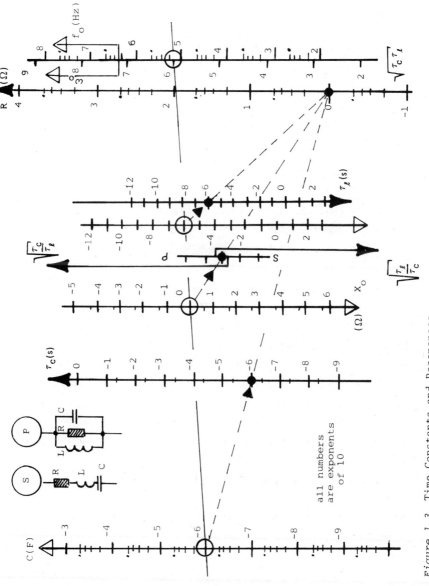

Figure 1.3 Time Constants and Resonances

with the roots of the ratio (Q) and of the product $(1/\omega_0)$ of the two time constants. Order of magnitude (OOM) calculations are facilitated by the nomograph of Figure 1.3.

It is hoped that the foregoing extractions--although highly concentrated--are sufficient to provide the reader with the very much necessary understanding of a basic tool he or she will need later for EM control: the reduction of amplitudes and rise times imparted by losses and frequency selection.

Inasmuch as we confined ourselves to linear circuits, we left out two important phenomena conditioned by nonlinearity: ferroresonance (see Ref. [1]) and subharmonics (see Ref. [2]).

1.1.1.2 Single Pulses

Long and Short Pulses

We saw that the double exponential pulse [Fig. 1.2 and Table 1.2, $F_1(Q)$ function for $Q < 0.5$] is the result of a step function. The step function and its deformation by neighboring circuit elements also suffices to describe rectangular pulses (and their modification) as the sequence of a positive and a negative step.

If the pulse duration is sufficiently long that the transient can die out as in Figure 1.4, then, for example, the $+f(Q)$ and $-f(Q)$ do not interact. [Note how easily we can apply the $f(Q)$ functions by simple inspection: If the steady state on-condition is not 0, the $f_3(Q)$ function is taken; if it is 0, the $F_1(Q)$ function is appropriate.] The Figure 1.4 response is the current generated by a ($R_i = 0$) voltage source or equivalently, for corresponding parallel resonance circuits, the voltage generated by a current source ($R_i = \infty$).

If the pulse is shorter than the time needed for the transient to die out, the difference for the shift T has to be taken as shown with conventional L-transform terms in Table 1.3, with Figure 1.5 giving the results for various pulse widths normalized such that $VT = 1$.

Useful Theory 13

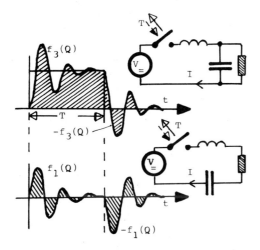

Figure 1.4 Long Pulses

Standard Pulse (1/50 μs)

An often used standard for a pulse is a 1 μs rise- 50 μs fall-time double exponential function. Now, since the rise (and fall) time is, by convention, defined as (1/0.8 times) the time interval between 10 and 90% of the pulse height, the 1/50 μs pulse actually looks as shown in Figure 1.6.

1/50 μs pulse: $V = 1.04(e^{-t/68 \text{ μs}} - e^{-t/0.39 \text{ μs}})$

Simpler 0/50 μs pulse: $V = e^{-t/72 \text{ μs}}$

The international standard (lightning) is 1.2/50 μs.

Pulse Spectra

Pulse spectra are so abundantly treated in textbooks that we describe here only a juxtaposition of some normalized pulse forms (Fig. 1.7) that are particularly instructive and useful.

Example 1.4 The spectrum of the raised cosine pulse is essentially confined to frequencies up to $f = 1/T$ (T as defined in Fig. 1.7!). That means: If we send a rectangular pulse of duration T through a low-pass filter with a cutoff frequency $f_c = 1/T$, the pulse is very much transformed--into a cosine pulse (a 1 kHz low-pass filter will convert a 1 ms pulse into a cosinusoidal pulse).

Table 1.3 Pulse Applied to RLC Series Circuit

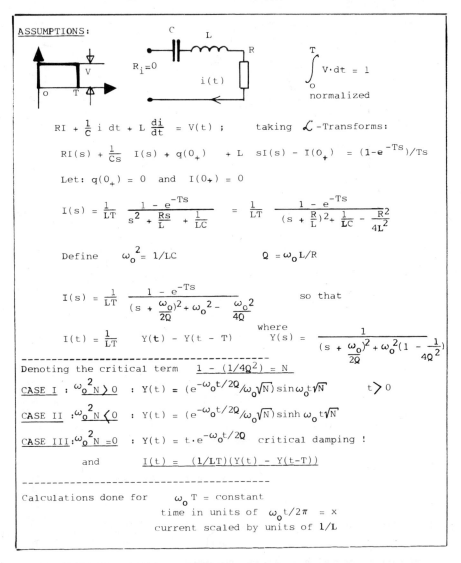

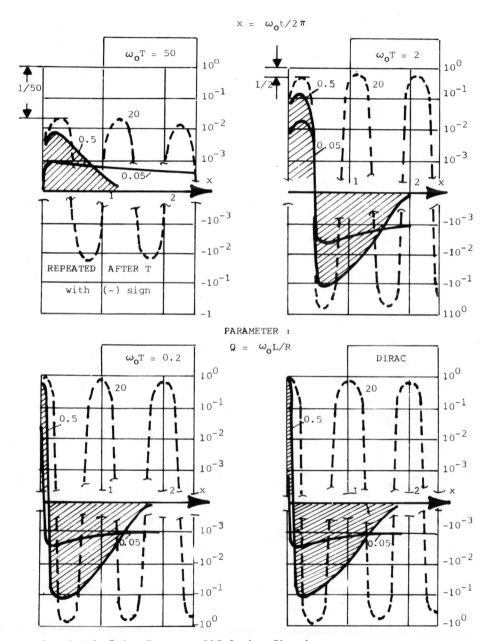

Figure 1.5 Pulse Response RLC Series Circuit

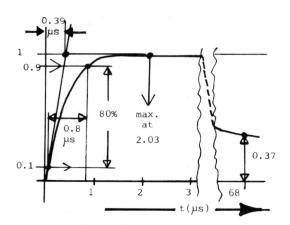

Figure 1.6 The 1/50 μs Pulse

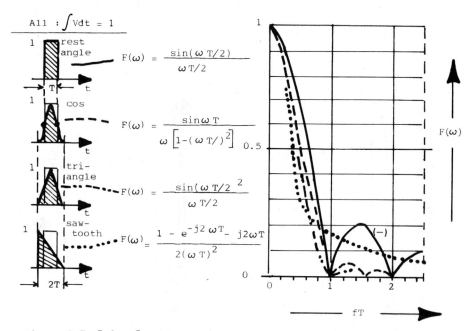

Figure 1.7 Pulse Spectra

It is also interesting to compare the two triangular pulses of the same base and height: The isosceles (= the squared rectangular pulse spectrum!) hardly contains energy above fT = 1. In contrast, for the sawtooth, which has the same area, the high-frequency content is not negligible.

1.1.2 Field Concentrations in Simple Geometries

To make our system operate safely, we must clearly understand how field concentrations come about, so that we can avoid their detrimental effects (danger of saturation or breakdown). In polarizable materials (high μ, high ε), saturation may transform the material into a very low μ or ε material, locally or over the whole body. Thus our counted-on shielding effect or capacitor value may disappear by saturation and we might as well have used wood in such places.

There is no magnetic breakdown, but there is always the danger of electrical breakdown. Breakdown of strong electric fields in ionizable or carbonizable materials may cause severe damage (faults). Yet, even without breakdown, electrical field gradients may be so large that precorona and corona discharges cause electrical interference and subsequent slow degradation of dielectrics. In this context, we have to mention also the dangerously great field gradients near "grounds" (step voltage can kill!).

We shall now show that--except for edge and microscopic surface effects (for both of which we shall resort to empirical work for expediency)--we can explain field concentrations with the simple theoretical models compiled in Tables 1.4 to 1.6. In doing this analysis of potential hazards, we must realize that we do it not for academic purposes (although with academic means). Rather, in so analyzing we must always have "control" in back of our mind such that we extract the key mechanisms and their critical parameters. We want to establish subcritical field conditions to prevent failures and hazards.

Table 1.4 Field Concentrations (Planes and Spheres)

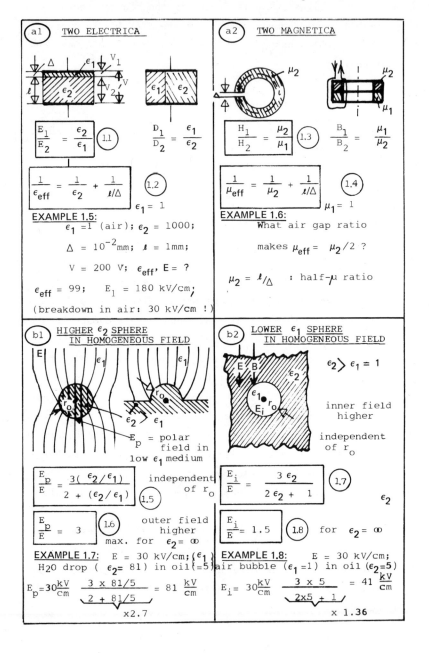

Table 1.5 Field Concentrations (Ellipsoids)

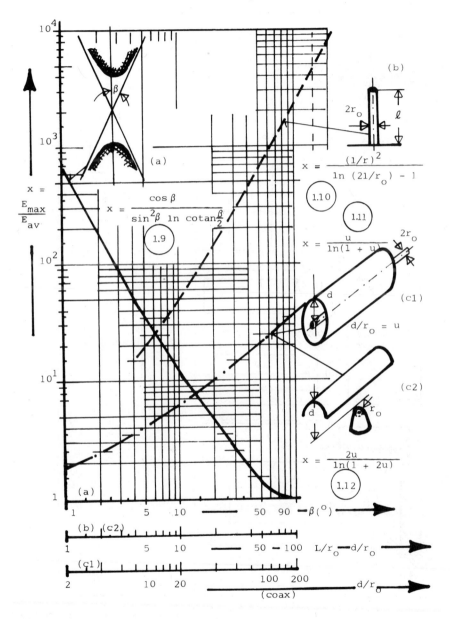

Table 1.6 Contacts (Grounding)

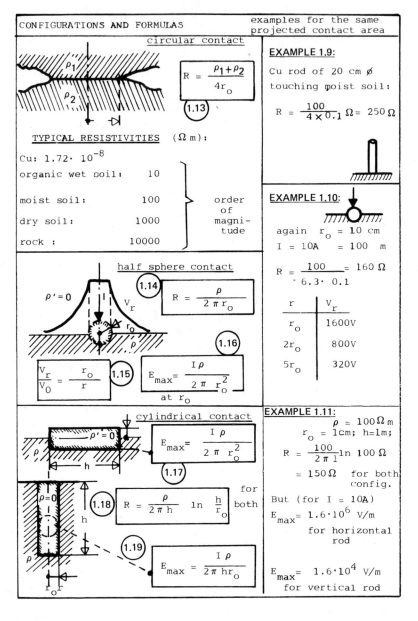

1.1.2.1 Two Materials in Essentially Planar Fields
Disregarding Fringing and Microscopic Effects

Table 1.4 is concerned with the combination of two dielectromagnetic materials, either two insulators, ε_1 and ε_2, or two magnetica, μ_1 and μ_2. Initially we consider only uniform fields, but make a clear distinction between the two materials in parallel (different fluxes) or in series (same flux through both materials).

For parallel materials (same E or H, respectively), there is no field enhancement. Only the dielectric flux densities D_1 and D_2, or correspondingly, the two magnetic flux densities B_1 and B_2, are just proportional to their corresponding material constants. Although there is no field enhancement in the dielectric case, the breakdown strength (withstand voltage) of one of the parallel materials may be substantially lower than that of the other. The breakdown of air in pinholes of dielectric films is a case in point. Later we make use of the seemingly odd property of two magnetic materials in parallel having a common winding (upper right corner of Table 1.4): the equivalent is two inductors (with μ_1 and μ_2, respectively) in series.

The situation is quite different for two materials in series (same flux through both materials): This is presented in Table 1.4 (a1) and (a2), the left side in each case. The field strengths are inversely proportional to the ε or μ ratios. Let us take (a1) in Table 1.4: a ceramic capacitor with not-fired-on electrodes imparting an air gap of $\Delta = 0.01$ mm. Thus, for example, for $\varepsilon_2 = 1000$ (ceramic) and $\varepsilon_1 = 1$ (air), we have a field strength 1000 times higher in the air gap than in the ceramic, or in our specific case, 180 kV/cm, which certainly will cause glow discharge. Hence high-ε ceramics must have their electrodes fired on to prevent such intolerable field concentrations. (By the way, excessive heat for too long during soldering may destroy the bond between metal and ceramic, or even the use of non-silver-bearing solder may rob the silber from a fired-on silver-glass-frit electrode, which is then destroyed.)

Air gaps also bring about a drastic reduction of the effective ε or μ [Eqs. (1.2) and (1.4)]. A simple relation to remember is that the effective ε or μ is half the harmonic mean of ε_2 (or μ_2) and a fictional ε_f (or μ_f) equal to ℓ/Δ. Thus in the case above, the seemingly negligible air gap of only 0.01 mm reduces the original dielectric constant of 1000 to an effective value of 99.

The great influence of air gaps is not limited to polarizable, nonlinear, dielectromagnetic material alone. Even alumina, Al_2O_3, an excellent insulator, often preferred because of its very good heat conductivity (only exceeded by beryllium oxide), with its dielectric constant of 9, can cause trouble, as we shall see when we discuss aluminim wiring.

Edge and Surface Effects

Considerations made thus far disregarded edge and surface effects. We shall take care of these aggravating conditions now but shall confine our presentation to the more critical electrical case. For the complex conditions encountered, we confine ourselves to experimental investigations (but in Section 1.1.2.2 we shall go back to surprisingly simple and effective theoretical modeling of tractable geometries).

Readers will remember from their school days the concentrations of equipotential lines at the electrode edges of capacitors. They will also recall that one can drastically reduce field gradients by shaping the dielectric in conformity with the equipotential lines (called the *Rogowski profile*). Such special shaping, with its increased costs, is justifiable for ceramic transmitter capacitors for high-voltage ratings because of a significant size reduction. But for conventional, mass-produced ceramic capacitors, such special stress-relieving shaping would be prohibitive. For foil capacitors, having low ε_2, the edge effect is less pronounced, and at any rate, the use of foils of inconstant thickness would be utterly impractical.

Let us study the field close to the thin and ragged edge of a metal electrode fired on a high-ε ceramic capacitor where three interfaces are involved: air/metal, air/ceramic, and ceramic/metal.

A good mathematician, after some effort, can make a computer print out field plots for various $\varepsilon_2/\varepsilon_1$ ratios. But knowing the indefiniteness of the edge, and consequently doubting the validity of the model, we prefer the experimental approach which in the course of our investigations will give us additional clues which we did not suspect at all. We start with an "electrolythic trough" using different surface resistance papers to plot the equipotential lines about the endgap of a bolt-mounted ceramic tubular capacitor. Figure 1.8 juxtaposes two versions: with and without a metallic end flange at the ceramic tube. (We selected only an $\varepsilon_2/\varepsilon_1 = 100$ due to lack of other resistance paper, but the trend is clear: $\varepsilon_2 \gg \varepsilon_1$.) If in Figure 1.8(a) the inner electrode is extended to an end flange, the field concentration close to the edge of the outer electrode is more than doubled compared to configuration (b), which does not have a metallic flange. (The flange pushes the field toward the edge.) This poses a dilemma: We ought not to have an end flange, then; but for mechanical reasons the face of the ceramic tube should be metallized. The way out: Use a metallized flange but no endgap on the ceramic tube.

While the macroscopic investigation just described has definite merit, it does not account for microscopic phenomena (the raggedness of the electrode, the roughness of the ceramic surface, and the presence of adsorbed air layers). Under the microscope, the area of interest appears as a big mess: very coarse and irregular, hardly describable by the simple geometries that we shall find so handy in Section 1.1.2.2.

Further complications arise from adsorbed air films trapped between the nonmetallized ceramic and the insulating potting compound, supposedly improving the breakdown voltage. Under sufficient stress, the tiny amounts of trapped air form microplasmas that generate interference and lead to progressive degradation and finally breakdown. The noise generated constitutes demonstrable damped oscillations excited by the negative V/I characteristic of the gas discharge in the microplasma. Figure 1.9 illustrates this

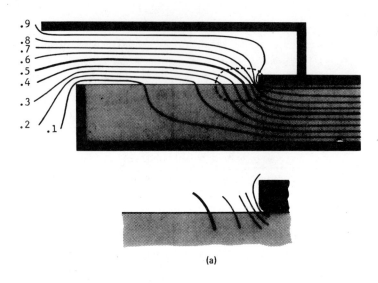

(a)

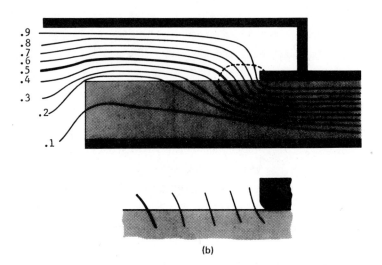

(b)

Figure 1.8 Equipotential Lines of Bolt-Mounted Ceramic Feed-Through Capacitor ($\varepsilon = 100$)

phenomenon, which is quite contrary to expectations (which are lower breakdown voltages at higher temperatures). Figure 1.9 presents the onset of the discharge voltage (noise!) and the breakdown voltage of

Useful Theory

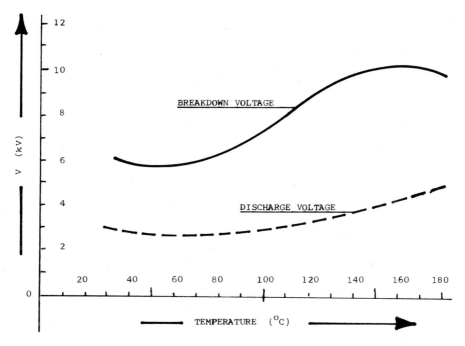

Figure 1.9 Unexpected Temperature Effects

uninsulated ceramic feedthrough capacitors immersed in transformer oil (a quality control test done before potting), as a function of increasing oil temperature. At first sight, it looks peculiar that both critical voltages should increase by 80% if the oil temperature is raised from room temperature to 150°C. A closer inspection reveals that heat creates tiny bubbles of adsorbed air on, and drives them off, the ceramic surface.

As Figure 1.10 shows, applying what we learned from the experiments leads to a significant improvement: By using a large endgap (field reduction) and surfactants in conjunction with the epoxy insulation (eliminating adsorbed air), the average value of the onset of the discharge voltage is improved by 300%. What is more, extrapolation of the straight lines in cumulative-percent probability paper (indicating only one failure mechanism) says that only 1 of 1 million improved filters will have a discharge voltage lower

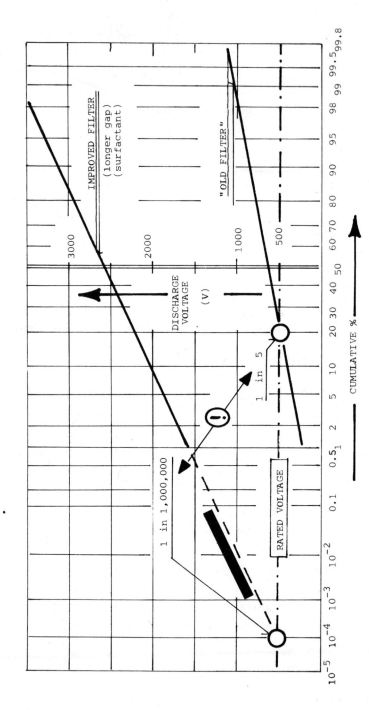

Figure 1.10 Improvement of Onset of Discharge Voltage

than the rated voltage of the filter; for the "old" filter, the ratio was 1 of 5. A significant improvement in interference and interference-related reliability has been achieved.

1.1.2.2 Easily Tractable Nonuniform Fields

We just discussed nonuniform fields having three different material interfaces and stated that such complex fields are best handled by carefully planned experiments. If we restrict ourselves to two material interfaces, we can approximate many practical situations quite closely with the theory of potentials. In Tables 1.4(b), 1.5, and 1.6, we have compiled the field concentrations occurring in such easily manageable geometries based on ellipsoidal coordinates.

Spheres

In Table 1.4(b) we juxtapose field concentration on (or in) spheres or half-spheres in an otherwise homogeneous field; (b1) relates to high-ε_2 (μ_2) spheres in a "thinner (lower ε_1 or μ_1) medium, and (b2) portrays the reversal of conditions. The examples indicated refer to (b1), water drops in oil (E_p is the field concentration at the pole of the sphere, coming close to the maximally possible factor of 3 for very high $\varepsilon_2/\varepsilon_1$), and for the opposite case, (b2), refer to air bubbles in oil (E_i is the field strength inside the air bubble). It can never exceed, even for very high ε_2, the factor 1.5. The field enhancement may be sufficient for ionization.

Note that the field concentrations under discussion are functions only of the geometry, not of the absolute size. Yet the absolute size plays a significant role for nonlinear relations: In Table 1.4(b), assume that the sphere is a thin shell of thickness d only of a high-μ_2 material. The sphere collects three times the flux that would pass through its projected area of $r_0^2 \cdot \pi$ in the uniform field. This flux, then, is further concentrated, equatorially, into the area $2\pi r_0 d$. For the same d, a large spherical shell can be saturated easily, whereas a small shell (r_0 small) will be much less affected by a static external magnetic field. We make

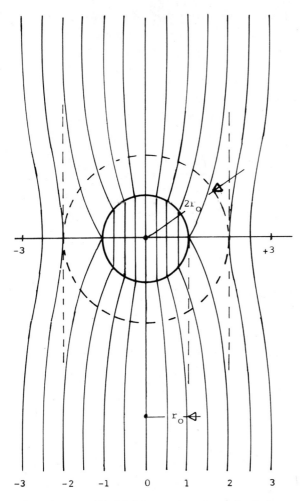

Figure 1.11 Field Distortion Near Cylinder

use of this important fact in Chapter 7. For a hollow sphere, the equatorial flux concentration is $3r_0/2d$. For a hollow cylinder, the 3 has to be replaced by a 2, as the exactly portrayed deformation of the external field of a cylinder (Figure 1.11) indicates.

Useful Theory

Metallic Cones and Spheres

Pointed configurations cause great localized field concentrations which we may wish to prevent or which, in some cases, we may wish to create at certain points for protection (lightning rods).

In Table 1.5(a) and (b), the factor $x = E_{max}/E_{av}$ is the field enhancement encountered by elongated metal structures in a uniform field E_{av}; Table 1.5(c1) and (c2) give x as it occurs at small curvatures inside cylindrical structures. -x for cones (a) and rods (b) is a measure of the protective value of lightning "protectors." Corona will reduce a very high x to lower values. But this should not disturb us once the discharge is initiated for protective purposes.

Contacts and Grounding (Single Point)

Field concentrations also play a key role in contacts, in particular in grounding systems. Table 1.6 (p. 20) lists the contact resistances and field concentrations in the immediate vicinity of the ground metal for approximations of practically important configurations. Examples 1.9, 1.10, and 1.11 refer to the same projected contact area and the same soil condition (moist soil). In view of the great variability of the soil resistance (time, depth, current, etc.), the differences imparted by the shape are roughly negligible. But this statement does not apply to the great field gradients occurring close to the metallic ground piece. Note the great differences in E_{max} (field strength in soil adjacent to metal) for the two options given in Example 1.11. Or in Example 1.10 observe that a current of only 10 A causes a voltage difference of 800 V between the metal and a point only 10 cm away from it. For lightning, let us say 10,000 A, this voltage would increase to 800 kV for the same 10 cm spacing. If we increase r_0 to 100 cm, the voltage drop 10 cm away from the ground metal would be only 8 kV.

The voltage gradient tapers quickly off with the distance from the "ground," but is often still high enough to kill an improperly located living beast (step voltage). But who can tell a cow to stand not too close and not radially (larger step voltage) to a tree into which a lightning may strike?

We shall recall the relations just discussed in Section 1.3.

1.2 REAL SOURCES

1.2.1 Criteria for Identification, Quantification, and Categorization

Most real sources of noise and hazards are more complicated and less predictable than the idealized theoretical ones discussed thus far. We come now to the formidable task of identifying, quantifying, and categorizing these real sources, which are very various and variant such that they seem to defy any attempt to formulate a practicable data base. How shall we best do this?

Let us start with identification (*what* attribute is critical?) and quantification (*how much* of it is critical?). This criticalness is determined solely by the susceptibility of the receptors involved. If we had no receptors (and no transfers), sources would be immaterial.

As we shall see in more detail later, quite often a pronounced conversion (not always easily predictable) takes place in the transfer from source to receptor (e.g., dV/dt becomes I, or normal mode is converted into common mode, etc.). We have to take this conversion into account, although, luckily most, but not all, transformations constitute a lessening of criticalness.

Here, then, are the primary (direct from source) and derived (converted in transfer) *critical attributes* of sources we have to know:

1. *Origin of electromagnetic interference (EMI)*. For control it is important to know where the interference originates (location, external, or internal to system).
2. *Frequency*. This may be a 10 Hz (±2 Hz) field, entraining the alpha waves of a person, causing disorientation, or it may be a harmonic of a radio transmitter causing RFI.

Real Sources

3. *Voltage and dV/dt*. Causing electrical breakdown or playing havoc in digital data handling.
4. *Current and dI/dt*. Excessive heat (and fires) are caused if too much current lasts too long or dangerous voltage spikes happen if currents change in too short a time.
5. *Impulse energy (plus pulse duration and rise time as modifying parameters)*. It may create destructive or interfering *resonances* or may stun or kill a person.
6. *Polarity*. This can be a decisive factor even for low-energy pulses. A wrong polarity of only 10 V can kill a transistor. Whether a pulse is additive or subtractive in terms of a bias voltage determines possibilities of suppression, and so on.
7. *Mode (normal or common)*. It often makes a difference whether or not a disturbance does any harm.

The *values of the critical attributes* of EMI range over many decades (entries marked with an asterisk should be watched in particular for 60 Hz and harmonics):

Frequency (operating): 0 to 10^{10} Hz*
Frequency (resonance): 10^1 to 10^9 Hz*
Voltage (amplitude): 10^{-6} to 10^6 V*
Voltage (change): 0 to 10^{11} V/s*
Voltage (field strength): 10^0 to 10^5 V/m*
Current (amplitude): 10^{-9} to 10^5 A
Current (change): 0 to 10^{10} A/s (this means dB/dt!)*
Current (field strength): 10^{-6} to 10^8 A/m*
Power: 10^{-9} to 10^9 W
Energy (of pulse): 10^{-9} to 10^7 J
With pulse width: 10^{-9} to 10^{-2} s ⎫
Pulse duration: 10^{-8} to 10 s ⎬ rise times
Mode: normal, common, mixed, direction of energy flow

For practicality, we shall categorize EMI (noise and hazard) sources according to broad classes of their (generating) *causes*:

Intentional, human-made sources
High-energy atmospheric transients
Switching transients
Material-caused sources
Limiting noise   ⎫
                      ⎬ very briefly only
Biological "noise"   ⎭

Admittedly, there may be better ways to classify EMI, at least from an academic point of view. But our classification is not made for academic purposes. And since the definer of any classificatory (ordering) system always has the choice of genera, we might as well classify this mishmash such that we facilitate the achievement of our primary goal: to categorize the multifarious EMI sources (by origin) to permit effective control. This is best done, it seems, by classifying the sources according to the mechanisms that cause them. We shall take a broad look and go beyond Ref. [4], which compiles the most critical switching transients identifiable in industrial control systems. However, even the carefully selected "Noise Guide" committee of the IEEE, IAS, was unable to establish a statistical data base on characteristics of sources encounterable in industrial control systems, much less in "any" system.

Throughout the text, we shall use the easily remembered acronym *FATTMESS (frequency, amplitude, time, temperature, mode, energy, size, statistics)* for the *critical attributes* we have to deal with in EMI/EMC situations. The attribute "size" should be understood rather broadly; that is, it includes structure, geometry, material, and circuit.

1.2.2 A Perspective Look at Real Sources

Information on specific aspects of EMI sources is widely dispersed in many a publication. A systematic statistical compilation of EMI sources is an elusive affair, except of indicating possible worst cases. Each system, even within the same class of systems, has its own resonance frequencies (ranging from kHz to hundreds of MHz) and its own Q's (mostly about 1, but sometimes about 30 or 40). Hence each system creates its own pulse shapes, which are only significantly different for exogenous disturbances.

We shall refer to some key references selected for their clarity of presenting the underlying mechanisms and critical parameters. Here we take only a bird's-eye view of the vast range of phenomena that may cause trouble.

1.2.2.1 Intentional (Narrowband) Sources

Intentional sources are comprised primarily of continuous or pulsed "single"-frequency transmitters as used for communications, radar, industrial or medical uses, and so on. They cover a frequency range of medium to microwave frequencies and may have field intensities of several 100 V/m in the neighborhood of large transmitters. In our context, we are not concerned with frequency management or utilization, which play a dominant role in communications systems. In fact, even precautionary site surveys (for EMC) or present FCC rules are of no avail, if we have, for example, the following situation. We have a smoothly running computer-controlled printing installation. One day a company using high-frequency pulses (HF) for sealing plastic bags moves in. As soon as they start operating, fully legitimately, our printing goes haywire. Retrofitting, by shielding and filtering, is the only way out since we did not buy controllers that were made immune to EMI from the very beginning.

1.2.2.2 High-Energy Atmospheric Transients

Two phenomena are involved:

1. *Lightning.* Lightning, a natural event, happens rather often. Normally, systems are protected against direct lightning (see Sec. 5.1). In 110/220 V circuits, lightning may introduce voltage spikes up to 5000 V, either critically damped or oscillatory, with fast decay with an oscillatory frequency somewhere in the kHz or MHz range. Reference [1] provides an excellent introduction to the physics of lightning and its interaction with power circuits.

2. *EMP*. EMP is a more energetic and more sharply rising, human-made phenomenon. It has to happen only once; and much of the sensitive electronic circuitry in the United States may be made inoperative unless EMP hardened as critical military installations are. To get an idea of the enormous energy involved, we indicate here only that an EMP wave may excite the wings of an aircraft ($\lambda/2$ dipole) with an initial peak current of 20,000 A $[f_1(Q)]$. In contrast to a nuclear explosion near the ground, a high-altitude nuclear burst, initiating war, may "illuminate" the continental United States with a pulse of over 10 kV/m, having a rise time on the order of nanoseconds. The reader is referred to Refs. [5] and [6].

1.2.2.3 Switching Transients

1. *Plasma physics and mechanical switching*. The complex phenomena of plasmas occurring in the switching of high power and their interaction with the circuits being switched are described in a tutorial article very much worth reading [7]. Reference [1], by definition, treats transient generation on a broader basis; and Ref. [4] gives various oscillograms of typical transients (including abnormal transients modified by current chopping) in 60 Hz industrial systems.

   The plasma created in mechanical switches is indestructible (but can be destructive for improper design), can be very hot, varies practically from 0 to ∞ in finite time [order of magnitude (OOM) 1 µs], and is a desirable thing to have in that it prevents di/dt from becoming ∞.

2. *Different phenomena in solid state switching*

   A. *Breakdown*. Here things are quite different. Heat, in particular localized heat, and carrier inertia are the limiting factors causing breakdown and/or interference. It is far beyond the scope of this discussion to describe in any detail the interaction of semiconductors and circuits. Most semiconductor manufacturers supply good application notes. But one point must be singled out. One must look for fast

recovery time to avoid destruction of rectifiers by transients. Chowdhuri [8a] points out that the transient-voltage-withstanding capability of silicon rectifiers may be lower than the steady state peak reverse voltage while carrying current in the forward direction. That is, if a device, passing a forward current, is suddenly biased in the opposite direction, its effective diode resistance remains momentarily low and briefly (1 to 50 μs) allows a reverse current burst to flow. (See also Sec. 6.1.)

B. *Glitches.* Here is a typical example of transient sources that may happen by inexact timing in logic circuits: called glitches. Two (remediable) causes make glitches appear at the analog output of digital-to-analog (D/A) converters:

   (1) Time skew between inputs (change input timing to equalize the arrival time of bits).

   (2) Unavoidably, unequal switching times inherent in saturated solid state switches (e.g., introduce a deglitcher or blanking window; see "When D/A Converter Glitches Rear Their Heads," D. Pinkowitz, Electronic Design, October 25, 1973).

1.2.2.4 Material-Caused Sources

Considerations thus far treated material effects in interaction with proximate circuits. Next, we take a brief look at those sources of problems that depend on material properties alone.

1. *Aluminum problems*

   A. *Aluminum wiring.* Aluminum wiring, on the basis of ampacity, is quite a bit more economical than copper wiring. In 1965, aluminum was introduced for branch wiring in quite a number of fixed and mobile homes. After some years, RF interference and fires occurred in such installations, in sufficient numbers to cause considerable alarm. Three mutually aggravating reasons were the cause:

(1) The "old" electrical grade aluminum had cold flow (copper has much less "give" under a screw head).

(2) Aluminum, exposed to air, immediately forms a thin layer of oxide, Al_2O_3, which is very hard, brittle, and highly insulating ($\varepsilon = 9!$).

(3) Aluminum has a much higher thermal expansion coefficient than copper or brass.

Altogether, with frequent heat cycling over many years, the screw connection gets loosened and arcing sets in.

In new installations, the price advantage of aluminum is no longer invalidated by such problems. Rather, new aluminum alloys, sacrificing only a few percentage points in electrical conductivity, but having very much reduced cold flow, are employed. With proper crimping tools, compression bonds, with metallic spikes penetrating the oxide layer, permit reliable connections.

In contrast, the "old" aluminum wiring is still a politically "hot potato." For some time, even a mandatory recall of such wiring was considered.

B. *Aluminum structures.* Typical are large reflector antennas. They are built by riveting, with aluminum rivets, aluminum sheets to supporting aluminum structures. Al-Al_2O_3-Al junctions, causing nonlinear tunneling, are formed (thin-film effects), which can cause spurious interference when irradiated with RF power (similar to the rusty-bolt effect aboard ships).

2. *Static electricity.* By the friction of different insulators, electrical charges, amounting to 20 kV or so, can be generated such that explosive materials are ignited or sensitive semiconductors destroyed (see Chapters 3 and 4).

3. *Corrosion.* A "normal," moist atmosphere can form the electrolyte of galvanic elements consisting of two different metals. This is the more pronounced, the more apart these two metals are in the galvanic series (given in many handbooks).

A supposedly well soldered ground connection may disappear within a few months by electrolytic action.
4. *Nonlinearity.* The nonlinearity of R's, L's, and C's is another source of trouble, as we saw in Section 1.1 and will see in more detail in Section 1.3.
5. *Mechanical vibrations.* Such vibrations, in particular in vehicles, may periodically open gaps that act like closed ones in the state of rest, such that a shield may become very poor for a moving vehicle.

1.2.2.5 Flak for the Amplification of Small Signals

1. *Intrinsic semiconductor noise.* Very small signals (OOM 1 µV) that must be amplified may be masked by the (frequency, amplitude, and temperature dependent) noise inherent in semiconducting devices (Johnson noise, shot noise, (1/f) noise) [9].
2. *Thermoelectricity.* Thermoelectricity caused in junctions of different metals may falsify small dc voltages.
3. *Muscle spasms (myographic pulses).* Muscle spasms may give a false indication (e.g., ringing in biofeedback amplifiers) [10].

1.3 COUNTERPRODUCTIVE EMC MEASURES

1.3.1 Heterogeneity of Objectives and Conditions Defies Simplistic Rules

It is often frustrating to the EMC neophyte that EMC measures backfire; that is, either they do not work or, if they solve a particular problem, they create a new problem. We shall now discuss--very candidly--the often all-too-human reasons for such counterpositive effects plaguing conventional EMC if applied in a formalistic way.

Certainly, it would be nice if we could formulate simple cookbook rules for EMC, rules that are not conditional, interactive, or even contrary in their objectives. But there are no simple, ironclad rules that can replace adaptive thinking if one strives for cost-effective electromagnetic compossibility of complex systems; too many facets are involved that cannot be taken out of context.

There is no argument regarding the need for guidelines that allow us to cope predictively with potential EMI and hazards. Under any circumstances we must abide by the NEC. Many standards, specifications, and codes do exist, in particular those established by the military services, the FCC, the FAA, and under the auspices of ANSI, the National Fire Protection Agency. The IEEE and various trade organizations, such as EIA, SAE, NEMA, and so on, are also very much involved in standards work. In the United States, Underwriter Laboratories, Inc. (UL), is the independent not-for-profit organization testing for public safety. On the international scene, CISPR and, to some extent, VDE standard prevail. Many of these standards are of considerable value, and quite a number of them are continually being updated in the light of new problems and findings. (For more details, see Chapters 5 and 10.)

But there are several big BUT's that severely limit the utility of standards in that they are, often unavoidably, conditioned on a delineated set of circumstances, thereby not addressing the whole problem on hand. In particular:

1. Some standards, to be simple and easily implementable, are based on such simplifying assumptions and arbitrary uncritical limits that the results of their application are, at best, misleading. Reliance on such unrealistic standards and specifications is dangerous in that it leads to a false sense of safety. Such highly imperfect specifications must be replaced.
2. By their very nature, standards and codes are general rules that may be--and very often actually are--perfectly correct for a delimited class of problems of a presumptively simplified reality. But the real problem is often much too complex to be tractable by a cut-and-dried approach. Thus the problem is often not the standard, but the engineer too lazy to think.
3. Then, there are the consequences of a poor, unimaginative engineering education. With a few exceptions, engineering education is education in engineering laws but not in engineering thinking. Textbooks, it is true, have to make simplifications and

idealizations as tentative guides. But the simplifying assumptions must not transform the real problem into one that becomes manageable but irrelevant. In EMI, even such simple concepts as an inductor supposedly being an "L" or a capacitor supposedly being a "C" must be questioned because of the vast range and scope of parameters encountered (FATTMESS; see Chapter 4).

Good standards are necessary but not sufficient conditions for EMC. We must approach EMC with a highly flexible and adaptive attitude, fully aware that a raw, unarranged problem facing the practicing engineer is never the same as an academic or bureaucratic exercise neatly preformulated.

Surely, in lawsuits, juries will be impressed by adherence to codes (authority!) and will decide against those who violated them. But juries will probably decide against a party--even though it proves strict adherence to a code--if it can be shown that the code cited is not applicable or insufficient (see the sailboat accident to be discussed shortly) in that the predigested code does not take into account a basic law of physics dominant in the case under consideration.

A good engineering boss will recognize and limit the functions of those incompetent engineers who substitute predigested formulas for thinking (try to solve nonroutine problems by routine methods).

Finally, there is one more "BUT," an absurd "but," a political "but," and very real. This "but" refers to the making of standards by committees. As problems grow, so do committees grow, with little coordination. Competent and incompetent members sit on committees where decisions are made on the basis of consensus and clout or deals. Hence we have:

4. It is never fully admitted, but is nevertheless true, that some standards are not established or updated as they could and should be. The change might be to the detriment of a special interest group: national prestige in conjunction with international standards, or the temporary economic advantage of

equipment developed under the old rules, or a buddy system trying to maintain a comfortable status quo. Bureaucracies and the NIH (not invented here) syndrome play a role. And in some cases, industrial representatives on committees withhold pertinent information lest they give up a competitive edge.

Thus, altogether, conventional, formalistic electromagnetic *compatibility* is often a rather imperfect thing, a compromise, not always based on the best information and facts available. But in time the facts will catch up.

But even if we were in the fortunate position to have perfect standards, they still could never automatically solve specific sets of problems (just as laws do not eliminate the judgment of specific cases by a judge). Rather, for cost-effective electromagnetic *compossibility* (see the Introduction) in the civilian domain, we have to analyze a system for coexisting conditions and their underlying principles and not by the partiality of a singular precept which may disregard a critical concomitant. But this is easier said than done. Since one learns only by doing, not by abstract theorizing, we shall now dissect several sets of simple, real-life EMC measures that are counterproductive in that they are sources of EMI or of costs. And we shall see, in preparation for later discussions (Chapters 4, 5, and 6), that electromagnetic compossibility is a thinking engineer's affair.

1.3.2 Illustrative Examples

1.3.2.1 Bureaucracy in Action

Before we discuss examples of a technical nature (that can be handled by reasoning), we shall bring just one example of an all-too-human nature (that defies reasoning because of the peculiar bureaucratic "mind" of the specification writer). In fact, we shall not even number this example because of its asininity, although it is very representative of rather common bureaucratic blunders and extravagances.

Example: The Golden Housing The author developed the antenna multiplexers for the Minuteman missiles (for in-flight data transmission of experimental birds). Extremely severe specifications had to be met, many of them justified. But in its eagerness to assure that nothing could go wrong, the contracting agency insisted adamantly on a number of excessive conditions that were completely unjustifiable but very costly. One point may suffice: Because some specification, somewhere written by another bureaucrat, stipulated it, they insisted on a very high reflectivity of the housing (aluminum). For permanence, gold plating was selected. But microscopic impurities in the aluminum caused pinholes in the gold layer, which in turn caused tiny blisters if subjected to salt-spray conditions. To avoid this, very purified aluminum had to be selected and meticulously tested before the interfacial layer of nickel was put on and after the goldplating was put on. Altogether, the "perfect" goldplating with its meticulous quality assurance procedures and its large reject ratio, was a costly affair. But alas, one day, the author had the opportunity to inspect the filters installed in the missile. They were shining and glimmering amidst gray and black devices having openings through which salt spray would have quickly destroyed vital control functions. But what is salt spray doing there anyhow?

After this unbelievable but true story, let us marshal a number of illustrative examples of counterproductive EMC measures imparted by the thoughtless application of standards and concepts. The examples are carefully selected for didactical purposes from such supposedly simple topics as filtering, shielding, and grounding/wiring.

1.3.2.2 Shielding Can Be Tricky Even at Low Frequencies

We shall see in Chapter 7 that good shielding for high frequencies requires careful consideration, in particular with respect to discontinuities such as holes and gaps and that good shielding for low frequencies is difficult because of the large skin depth involved. Here, in contrast, we point out two important but often

Table 1.7 Double Grounding

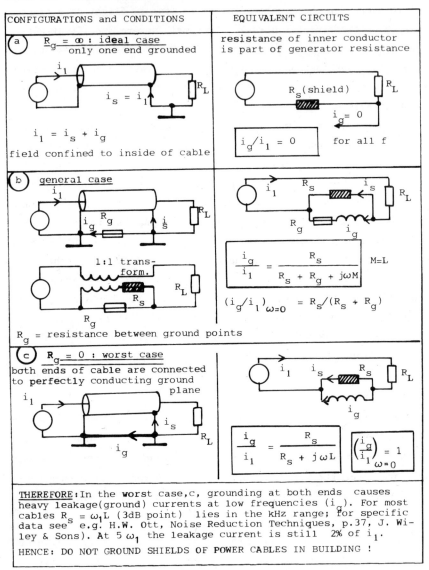

overlooked shielding phenomena causing seemingly perplexing problems at low frequencies simply by the very way the shield is connected:

1. *Cable leakage current by double grounding* (Table 1.7). The "peculiar" phenomenon that a cable grounded at both ends leaks current into the grounding path--the more so the lower the frequency--can easily be explained by means of the equivalent circuit of Table 1.7(b). In the extreme case of Table 1.7(c), the perfectly conducting ground plane, all the return current is leakage current at dc. As the transformer action of the cable increases with frequency, the current is more and more confined to the cable at higher frequencies [hereby we assume tacitly that the cable is not braided or has holes providing leakage at high frequencies (see Table 2.3)]. For most coax cables, the 3 dB point ($R_g = \omega_1 L$) is reached in the kHz range. Unsuspected 60 Hz stray currents are often caused by double grounding of cables (see the bottom of Table 1.7).

2. *Common mode/normal mode conversion, in particular at the all-pervasive 60 Hz power frequency* (Table 1.8). Let us assume that we have carefully twisted a pair of wires, leading from a sensor to a differential preamplifier, to prevent "any" magnetic pickup by reducing the effective area between the wires to zero (normal mode). Yet for the typical conditions presented in Table 1.8, we can run into unsuspected difficulties of varying severity, depending on how we hook up shields or filters to "prevent" (?) ground-loop problems. Since the mode conversion involved is a very basic EMC problem, we shall discuss it in some detail. Let us start with Table 1.8(a): The signal $V_s = 10^{-3}$ V is feeding into the load resistor R_L (input of amplifier) via the source (or inbalance) resistor R_s = 265 ohm. The return current, in the lower wire, is of opposite direction. This is called *normal mode* operation (also longitudinal or series mode). In contrast, a *common mode* signal of 10 V of the ubiquitous 60 Hz is introduced into the ground loop formed by the grounded source and the stray capacitances of the R_L via the twisted pair in which the 60 Hz current now flows in the same direction (parallel or common or transversal mode).

Table 1.8 Common Mode/Normal Mode Conversion

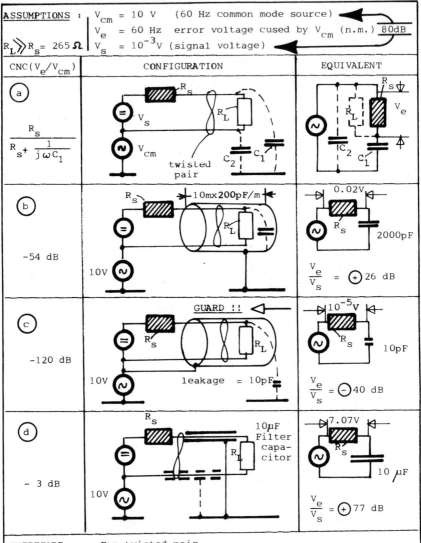

THEREFORE : For twisted pair
* Use guard connection and minimize residual (leakage) capacitance!
* Do <u>NOT</u> use filter capacitor (worst thing you can do!)!!!

With V_{cm}/V_s = 80 dB, even a small fraction of the 60 Hz voltage across the (normal mode) R_L can cause a formidable error voltage V_e (mode conversion). Table 1.8(b), (c), and (d) are instructive examples of what should be done (or not be done) to arrive at an acceptable 60 Hz error voltage. In (b), a 10 m cable shield (with 200 pF/m) is grounded at the load side. As the equivalent circuit indicates, the error voltage is now 26 dB above the signal voltage. If on the other hand, the cable shield is grounded as shown in (c), namely as a guard, and if we are careful such that only a very small residual leakage C (let's say 10 pF) is left at the load side, the error voltage is 40 dB below the signal voltage. Finally, in (d), we make a big blunder in trying to do good: A 10 µF filter capacitor raises the error voltage to 77 dB above the desirable signal voltage. Moral: Filters in the wrong place are really wrong.

1.3.2.3 Conventional Filter Books Are of No Avail

In contrast to the well-established and effective theory and practice of filters used in communications systems (based on impedance match), the theory and practice of filtering for EMC is still in an unbelievably poor shape (mostly because of variant and undefined mismatch). The ineffectiveness of power feedline filters is all the more consequential inasmuch as the lines permeate the whole system--as veins permeate the whole human body--and distribute interference throughout the system--as veins do with cancer. In Chapter 8 we attack and solve this long-standing EMC problem very fundamentally. Here, in our consideration of counterpositive EMC measures, we only highlight the problem of filtering for EMC, which, unless properly handled, does more harm than good and is one of the reasons why conventional EMC is held in such low esteem in some quarters.

More or less, all the reasons given in Section 1.3.1 account for the sad state of EMC filtering, but two major technical difficulties can be singled out as primary culprits. But they can be managed once their idiosyncrasies are clearly recognized:

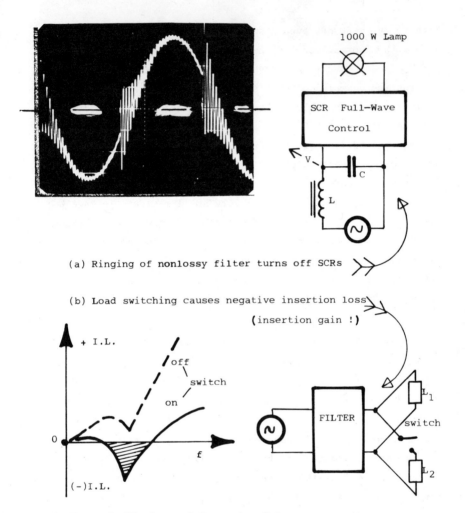

Figure 1.12 Ringing and Insertion Gain

1. *The indeterminate and variant mismatch (filter as a whole)* (see Fig. 1.12). The assumption underlying all "accepted" filter theory is impedance matching (Q = 1). This criterion is invalidated by power feed lines and the impedances of the power generators and loads, all of which are designed solely for good efficiency at the power frequency--nothing else. Two sets of highly undesirable effects can happen because of indeterminate and variant mismatch:

A. In the passband and the transition band (between the passband and stopband) ringing and/or insertion gain may occur. We met both effects in Table 1.1. Now Figure 1.12(a) portrays the detrimental effect of ringing. The excessive negative swing of the ringing prevents the SCR bridge from turning on the lamp. Introducing losses (e.g., in the form of thick laminations of the filter inductor) eliminates this problem (damping).

Figure 1.12(b) shows the consequences of insertion gain as it may be caused by switching off or on loads, thus changing the interface conditions of the filter. If, for instance, a harmonic has an amplitude of 10 V and coincides with the self- or eigenresonance of the filter having an insertion gain of 30 dB, the harmonic will be enhanced to 316 V amplitude, which could be destructive for the filter capacitor.

B. In the stopband, as Figure 1.12(b) also indicates, loading or unloading of the filter may drastically change the filter effectiveness. Refer to Chapter 8 for the solution of these heretofore unsolved problems.

2. *The large ranges of frequencies, amplitudes, and temperatures* (FAT; see later) (filter *elements* themselves do not behave as labeled). See Table 1.9, which compiles the most common pitfalls encounterable in *individual* filter elements. The label of the filter element is valid only over a limited range of frequencies (F), amplitudes (A), and temperatures (T). In FAT, we meet the beginning of the acronym FATTMESS (frequency, amplitude, temperature, time, mode, energy, space, and statistics), which we use throughout the text as a memory crutch referring to operating parameters to be considered for criticalness or modification.

Now, in Table 1.9(a), we see that what is supposed to be an "L," after passing through a parallel resonance, turns into a "C" at higher frequencies (distributed capacitance becomes dominant). Corrective steps are indicated.

Table 1.9 Nonideality of Filter Elements

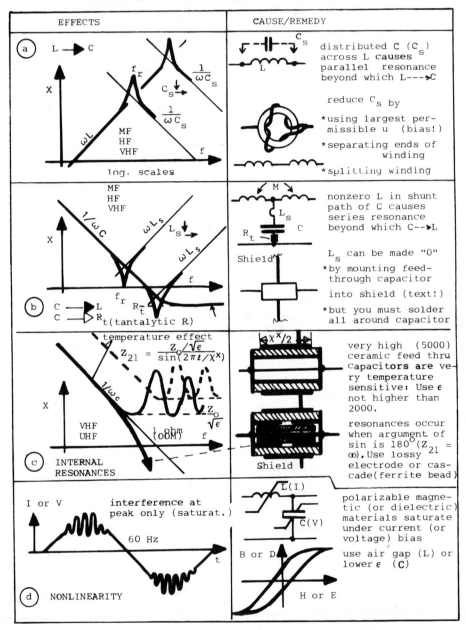

Conversely, in Table 1.9(b), a thing labeled "C" may turn into an "L" at higher frequencies, after passing through a series resonance. This L is the opposite of what is desired in an expected HF short-circuiting shunt. It is caused by the lead wires of the capacitor. In fact, even if we use a so-called feed-through capacitor, supposedly eliminating any M and L_s, we must make certain to solder the feed-through capacitor into the shield all around its periphery lest we leave a little gap or hole which would constitute a noninfinite number of parallel L's, and hence not L = 0 in the shunt path.

In the case of tantalytic capacitors, a residual R_t, in the order of 1 ohm, prevents the capacitor from behaving like a capacitor at high frequencies. We must, therefore, always shunt a ceramic feed-through capacitor in parallel to assure good high-frequency behavior.

In Table 1.9(c) we are concerned with a phenomenon that shows up at very high frequency (VHF) and ultrahigh frequency (UHF), even if we use ideal mounting as just described. Internal resonances are responsible for the resonance peaks shown.

Example 1.12 [*Table 1.9(c)*] Assume that we have chosen a relatively low dielectric constant ε of 2500 (to minimize the temperature effect indicated in the diagram) for a tubular ceramic feed-through capacitor having an effective electrode length of 1 cm. We must realize that the propagation (TEM mode) in this electrically very dense material is very slowed down, as the first resonance occurs at the tubular structure constituting $\lambda^*/2$. The wavelength in air would then be 2 × 1 cm × $\sqrt{2500}$ = 100 cm or a resonance frequency of 3×10^{10} cm/s per 100 cm = 300 MHz. In air, the characteristic impedance of the tubular structure, which has a diameter ratio of 1.5:1 is about 25 ohm. The dielectric reduces this to $25/\sqrt{2500}$ = 0.5 ohm. The first resonance peak, for Q = 15, is $(2/\pi)Q$ or 5 ohm; whereas the (transfer) impedance of the ideal capacitor would have been the impedance of 3000 pF or 0.18 ohm. Luckily, such "suck-out" points can be eliminated. In fact, by properly incorporating ferrite beads, such ceramic feed-through capacitors can be made to behave better--actually much better than ideal capacitors (see U.S. patents 3,023,383 and 3,035,237, and problem 55 in Chapter 12).

Table 1.9(d) is concerned with the "A" of the FATTMESS term (amplitude). For polarizable materials (high μ or high ε), high amplitudes of I (or V) may saturate the material, thereby reducing the μ (or ε) to such low values that the filter function is impaired for high amplitudes, as indicated for 60 Hz bias. Filter measurements have to take such "modulation" into account.

1.3.2.4 Grounding Is Not Grounding

Here we refer to "safety" ground, not high-frequency or signal ground, in part discussed before. Careful thinking is necessary when grounding for safety must be considered. Although the underlying principles are simple, they are often grossly misunderstood, even by those who are supposed to know them and misapplied by those who teach them.

The energy contained in a lightning stroke or a fault current is so great that it can only be diverted into a "sink." But the immediate neighborhood of the sink entry often exhibits highly dangerous voltage gradients. Equalizing potentials (by providing large entry cross section and/or bonding) becomes necessary.

Table 1.10 sets forth three illustrative examples taken from real-life situations with the essence carefully extracted. The first two examples are based on Table 1.6 (grounding contacts) and the voltage cone depicted for the half-sphere electrode. The closer we are to the point of entry, the greater is the surface voltage gradient imparted by the energy diverging into the ground mass.

In Example 1.13 (the sailboat accident), Table 1.10(a), a man let his fiberglass sailboat, with his family aboard, glide into a bay. Suddenly, the aluminum mast struck a 69 kV high tension line crossing the bay. Although many fish were killed, great luck prevailed in that all people survived. The man was picked out of the water, in which he floated face down. His wife and daughter were badly burned. The accident was aggravated by the explosion of the

Counterproductive EMC Measures

Table 10.1 Grounding and Wiring for Safety

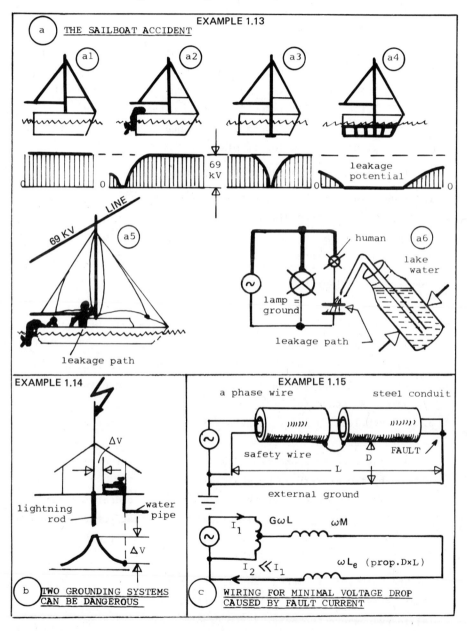

gas can on board. The boat was not "grounded" by a 30 cm × 30 cm (1 ft^2 in the Babylonian language) copper plate, as suggested by the "appropriate" yachting code for lightning protection. Lawyers for the Electric Power Company, sued for $1,500,000, argued that no damage could have occurred if the code had not been "violated." Their two expert witnesses, electrical engineering professors from a highly respected university, testified elaborately under oath that nothing could have happened to the people involved if the boat had been grounded via a copper plate.

It is not said that the professors were

1. Stupid in that they did not understand the situation
2. Naive in that they did not know that water is conductive and sprays
3. Lieing in the interest of their clients

But it is said that the professors

4. Tried to impress the justice and jurors with an elaborate display of neat-looking sailboat models floating in big tub. They demonstrated that under the conditions they created (very clean and dry surfaces of the boat models), arcing occurred across the model's surface when no "ground" plate, properly scaled down, was connected.
5. Nevertheless failed to convince jurors and judge for two unrefutable arguments based on facts:
   A. The counterarguments and demonstrations, given under oath by the writer, as we shall discuss shortly.
   B. The documented fact that in a similar accident, at much lower voltage (12 kV line) and with a "perfectly grounded" boat, a man was severely burned and is still suffering now, years after the accident.

The author's arguments and simple demonstrations regarding Item 5A are sketched in Table 1.10(a). Let us first look at the top line (a1 to a4), where the heavy lines denote a high potential caused by the contact of mast and power line. In (a1) the boat is not "grounded" and a high potential exists from the mast, boom, and stays (and all the people in contact with them) to the water surface.

In (a2) an arc to the motor or motor mounting plate causes a dip to "zero" leakage potential (to the water surface) in the immediate neighborhood where the metal touches the water. But not far away the leakage potential is again essentially the full line voltage.

In (a3) the mast is "grounded" by a "prescribed" 30 cm × 30 cm copper plate. Now the trough of the leakage potential is shifted to the immediate neighborhood of the "ground" plate (see Table 1.6). But anybody contacting any of the interconnected metal parts on top of the boat and having his or her feet a bit away from the trough will be exposed to nearly the full voltage toward the water surface.

In (a4) we suggest a "grounding" metal grid (preferably strips) on the boat to provide a large entry area and correspondingly a much broader potential trough, approximately a metallic boat. The danger of a hazard is thereby very much reduced.

After these preliminaries, let us return to the accident proper (a5). A person standing on deck and touching the mast, stay, or boom has his or her feet essentially at full line voltage against the water line, and any leakage to the water line (by dirt or water spray at the boat's surface) can cause an appreciable current through the human body.

Now consider (a6), the demonstration that, in conjunction with the 12 kV case, clinched the case in favor of the author's client, the insurance company of the boat's owner. A 100 W lamp was mounted on an insulating board and fed by 120 V ac. The lamp heating represents the energy discharged into the water. A little Christmas tree lamp, substituting for the person standing on deck and touching or being close to high potential, was shunted in parallel to the big lamp. But a gap between two metal strips (fastened onto the insulating board) was put in series with the little lamp and represented the leakage path from the person to the water line, scaled down by the ratio 69,000:120. Initially, the little lamp was not lit when the big lamp was lit. But upon spraying lake water (100 ohm-m resistivity) into the gap, the little lamp lit up. Thus this simple model proved the principal contention that grounding, as defined by the code, could not have prevented the accident. It was a pleasure to work with highly alert lawyers who might have lost the $1,500,000 case if they had listened to a cookbook expert.

After the foregoing detailed discussion, Example 1.14 ("side flashes"), Table 1.10(b), can be treated quite succinctly: If the wire between a lightning rod and its ground rod passes relatively closely to a metal object connected to another ground (e.g., water pipes, being on ground zero potential, arcing, or induction into the separately grounded object, let us say a bath tub, may cause severe hazards. The need for bonding, to establish equalization of potentials, becomes quite obvious. Again our reasoning is based on the voltage cone described first in Table 1.6.

In Example 1.15, Table 1.10(c), we stress the importance of proper grounding and safety wiring (*green wire*) in conjunction with steel conduits such that the voltage difference created by ground fault currents is minimized, an important consideration of particular significance in hospitals (see Section 5.3). Inasmuch as an inductance is given by $L = \mu\mu_0 AN^2/\ell$, the external spread-about ground L_e is determined by the area. $A = "L" \times D$ and constitutes a rather large ωL_e, resulting in a large voltage drop if it is the only ground. Yet this voltage drop can be very much reduced if we use the (green) safety wire inside the conduit *and* the conduit itself as a transformer with a bifilar winding such that only its leakage reactance is effective.

This concludes our selective discussion of counterpositive anti-interference measures which we categorized as unsuspected interference and hazard sources, although they are mostly matters of transfer. All the more, then, are we prepared for the next chapter: transfer, T, which looks at the many specific transfers possible. Only in Chapters 4 and 5--after much interactive analysis and control--will we be able to coordinate generically the intermeshing noise transfers complicated by mandatory safety measures and unavoidable spread of the system.

REFERENCES

1. Electrical Transients in Power Systems, A. Greenwood, Wiley-Interscience, New York, 1971.
2. Forced Oscillations in Non-linear Systems, C. Hayashi, Dai Nippon, Tokyo, 1953.
3. Potential Felder der Elektrotechnik, F. Ollendorf, Springer-Verlag, Berlin, 1932.
4. Guide for the Installation of Electrical Equipment to Minimize Electrical Noise Inputs to Controllers from External Sources ("Noise Guide"), IEEE Publication P518/D5, Industrial Control Systems Subcommittee of the IEEE, Industry Applications Society, New York, 1977. (2nd ed., IEEE Std. 518-82, in preparation.)
5. EMP Engineering and Design Principles, Bell Telephone Laboratories, Electrical Protection Department, Loop Transmission Division, Technical Publishing Department, Whippany, N.J., 1975.
6. EMP Radiation and Protection Techniques, L. W. Ricketts, J. E. Bridges, and J. Miletta, Wiley-Interscience, New York, 1976.
7. Circuit Breakers, I: Fundamentals, W. Rieder, IEEE Spectrum, July, 1970.
8a. Transient-Voltage Characteristics of Silicon Power Rectifiers, P. Chowdhuri, IEEE Transactions IAS, September/October, 1973.
8b. Electrical Interference from Thyristor-Controlled DC Propulsion System of a Transit Car, P. Chowdhuri and D. F. Williamson, IEEE Transactions, IAS, December, 1977.
9. Noise Reduction Techniques in Electronic Systems, H. W. Ott, Wiley-Interscience, New York, 1976.
10. Muscles Alive, J. V. Basmajian, Williams and Wilkins, Baltimore, 1967.

2 Transfers
Start with Definable Coupling

In contrast to communications systems with their clearly definable channels, non-EMC-planned control systems are characterized by a rather untransparent diffusion of interference from many broadband sources. Similarly, large control systems, spreading over large areas, contain 60 Hz ground loops and ground resistance coupling. Thus, instead of speaking of channels, we refer to them more appropriately as couplings or transfers. Because of the close proximity of sources and receptors, the very large ranges of critical parameters involved, unavoidable imperfections of shielding and grounding, and the all-pervasive penetration of branching power feed lines, the transfer, in particular of unwanted signals, is rather unpredictively spread about. Since lines are essentially all-pass networks, the spreading around of interference is so diffuse that attempts to calculate them exactly would be an interesting mathematical exercise without much practical significance. Hence it is imperative:

To understand thoroughly the principles and criteria of different specific transfers (as we shall do now)*

---

*See also spread transfers (Chapters 4 and 5).

To plan, by grouping, isolation, and so on, the system from the very beginning such that we do not build a system with undeterminable transfer and hazard potential (as we shall do in Chapters 4 and 5).

We are skipping here the well-known formulas for resistive, inductive, and capacitive coupling (see, e.g., Ref. [1]). Luckily, the basic relations we are looking for turn out to be rather simple, though not always familiar concepts. Numerical examples will help overcoming this unfamiliarity.

We are already partly prepared for the consideration of transfer problems by Sections 1.1.2 (fields) and 1.3 (circuits and fields). There we looked at field concentrations or circuit phenomena--both actually being transfer occurrences--as virtual sources or source modifiers of interference or hazards. In this chapter we are interested in viewing transfer as more or less independent of the source (from a broader point of view, in terms of EMC).

2.1 CONDUCTIVE TRANSFER

Both for conductive and radiative transfer, we must consider the frequency *and* the time domain to get a reasonable handle on analysis (and control) of transfer problems.

2.1.1 Frequency Domain

For practical purposes, we differentiate between conductive and radiative transfer. Yet we must be fully aware of the fact that even for conduction the energy is contained in the field surrounding the conductor. Thus, for both kinds of transfer, the key properties of propagation are significantly determined by the material around or between conductors. Material and geometry are the generic parameters.

2.1.1.1 Uniform Transmission Lines

Table 2.1 compiles the most important equations for constant Z_c transmission lines as far as we need them for EMC. With

Table 2.1 Transmission Line Equations

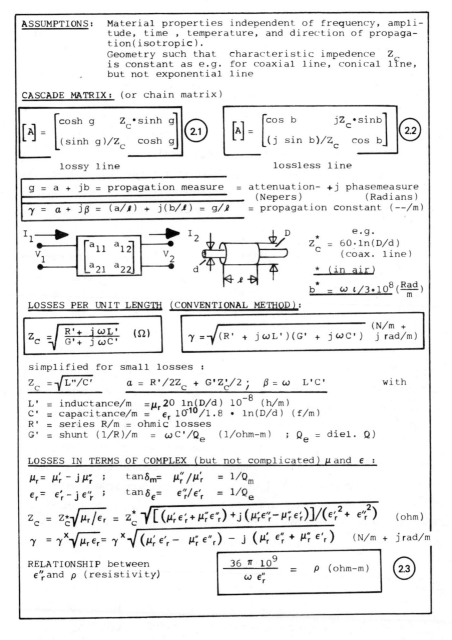

Equations 2.1 and 2.2, we juxtapose the cascade matrices of the lossy and lossless transmission lines.

One important advantage of the equations is that they greatly facilitate the derivation of lumped circuit equivalents of lines (wires and cables whose length approaches the $\lambda/2$ dimension).

Example 2.1 Refer to Example 1.12 and prove that the parallel resonances of tubular feed-through capacitors occur at n = 1, 2, 3, ..., $\lambda/2$ (300, 600, 900 MHz). The feed-through operates essentially into an open circuit. Hence $I_2 = 0$ gives $V_2/I_1 = Z_{21} = 1/a_{21} = -jZ_c/\sin b$ = transfer impedance of the equivalent T network (not to be equated with the transfer impedance/m = z_T, discussed in Section 1.1.3). For $b = \pi$, $Z_{21} = \infty$. With $b^* = \omega\ell/3.10^8$ and $\varepsilon' = 2500$, $b = b^*\sqrt{\varepsilon'}$ becomes π. q.e.d.

Let us examine this transfer impedance a bit more. The minimum of Z_{21}, even for the lossless case, is not zero but Z_c (or 0.5 ohm in the example), occurring at all $\sin b = 1$. Now, according to Foster's reactance theorem, lossless impedances of one-ports alternate between 0 and $\pm\infty$. Why, then, does this not apply here? Z_{21} is fictitious and as such not directly accessible. It is nevertheless a highly useful concept inasmuch as the series impedance of the equivalent T network is negligible in terms of the series R or L purposively inserted to make the C into an effective low-pass filter.

In Table 2.1, in addition to the conventional characterization of transmission lines, we introduce complex permeability and complex dielectric constant. These complex terms may seem clumsy to the uninitiated, but they are actually highly convenient. Here we give only a demonstration of the meaning of the imaginary part of permeability (see Equation 2.3 for the corresponding dielectric term, of which we make ample use later).

Example 2.2 A ferrite bead has an outer diameter of 0.5 cm, an inner diameter of 0.25 cm, and a length of 1 cm. What is its impedance at 150 MHz, where $\mu_r' = 100$ and $\mu_r'' = 400$? According to

Table 2.1: $L = \mu_r \cdot 20 \cdot \ln 2 \cdot 10^{-8}$ (h/m) $\cdot 0.01 = 13.9 \cdot 10^{-10} \mu_r$.
That is, $j\omega L = j \cdot 10 \cdot 13.9 \cdot 10^{-10} \cdot (100 - j400) = j13.9$ ohm + 55.6 ohm. That shows how the -j term of the permeability multiplied by the j operator of the impedance results in an equivalent resistor ($-jj = 1$). By the way, for many ferrites the μ_r'' is proportional to 1/f for frequencies above 1 MHz such that a ferrite bead at higher frequencies is essentially a constant RF resistor with some slightly inducive component, but no dc voltage drop happens.

2.1.1.2 Efficiency of Lines

Uniform Mismatched Lines; General Relations

For linear conditions, a source with an open circuit voltage V_{oc} and a short circuit current I_{sc} will provide the maximally available power into a load resistance of $Z_L = V_{oc}/I_{sc}$. This maximum of extractable power is $V_{oc} I_{sc}/4$, the efficiency 50%. But for high efficiency, approximating 100%, a voltage source must be operated close to open circuit, a current source close to short circuit conditions.

If we insert a transmission line between the source and the load, its characteristic impedance should be equal to source and load impedances for maximal energy transfer. Then, there are no reflected waves traveling back and forth on the line; and for a lossless line, the efficiency is 100%. For a lossy line, matched as stated, the attenuation concept is appropriate.

Example 2.3 Attenuation. Given a load resistance of 20 ohm, we select a transmission line of characteristic impedance $Z_c = 20$ ohm (or use a corresponding transformer). For a line resistance of $R' = 0.1$ ohm/m, what is the attenuation for (a) a 10 m line; (b) a 200 m line? According to Table 2.1, $\alpha = R'/2Z_c = 0.0025$ N/m. Hence (a) $V_0/V_i = e^{-0.025N} = 0.95$; (b) $V_0/V_i = e^{-0.5N} = 0.37$. What is the efficiency?

In many practical cases, matched conditions are not economically feasible; then efficiencies are much less than those calculated by attenuation. It must be kept in mind that the efficiency of a line

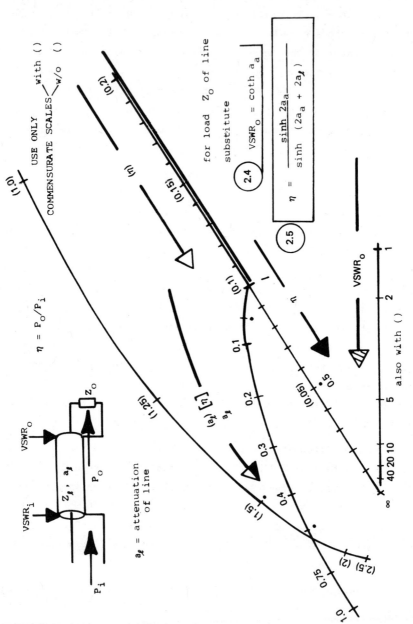

Figure 2.1 Transmission Line Efficiency

proper (energy delivered by the line divided by energy put into the line) is determined by the load mismatch *only*. High-frequency engineers use the *Smith diagram* to calculate the transmission line effects (impedance transformation and efficiency of mismatch). Here we make use of the fact that the Smith diagram is actually a conformal presentation of the hyperbolic cotangent with complex argument $g = a + jb$. The load $R_0 + jX_0$ creates a voltage standing wave ratio, $VSWR_0$, at the load: Equation 2.4, Figure 2.1. $VSWR_0 = \coth a_a$ means that we replace the load by an equivalent open line with the same Z_c and the attenuation a_a. If we have no Smith diagram, we may proceed as follows:

$$VSWR_0 = \frac{Z_c}{R_0} + \frac{X_0^2}{Z_c R_0}$$

The nomograph Figure 2.1 permits easy determination of the efficiency of mismatched lines (Eq. 2.5). It provides a good feel for the great reduction of η by mismatch.

There is an additional reduction of energy reaching the load; Figure 2.2 yields the multiplier of the line efficiency because of input impedance mismatch: the more lossy the line, the more the $VSWR_i$ (input) is reduced. The standing waves (or mismatch) become less toward the generator; in fact, for high losses the $VSWR_i$ becomes 1.

Two Important Extremes

Table 2.2 presents two illustrative examples (2.4 and 2.5) of extreme but rather common cases. They are self-explanatory:

1. Capacitive load and short lossless line (short cable feeding short antenna; short in terms of λ)
2. Very lossy line (lossy ignition cable and effective length)

Losses of Power Feed Lines

No direct data on the attenuation of power feed lines have been published (as far as the author knows). But from open and short circuit measurements on British aircraft cables rated 6 to 200 A

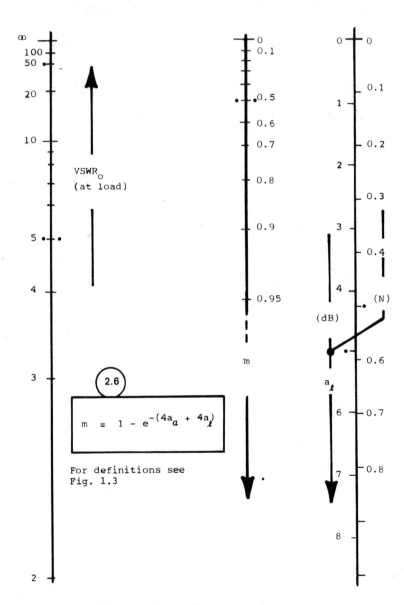

Figure 2.2 Effect of Input Mismatch

Table 2.2 Two Extreme Cases of Line Effects

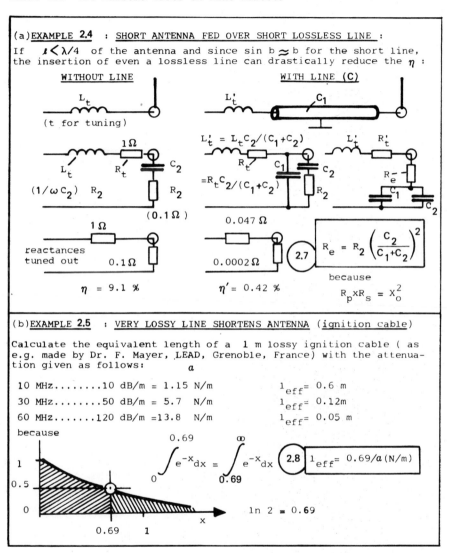

(published by the Royal Aircraft Establishment), we can calculate (how?) and use the following as an indication of the attenuation of power feed lines:

Upper limit: 0.0150 N/m
Mean value: 0.0050 N/m $\Big\} \times f(\text{MHz})/1\ \text{MHz}$
Lower limit: 0.0025 N/m

Apparently, for rather densely packed cables the losses are about equally distributed between eddy current losses ($f^{0.5}$) and dielectric losses.

2.1.1.3 Transfer Impedance of Coaxial Cables

This discussion refers to coaxial cables, but is also transferable to cable bundles in cylindrical shields or cable trays. In any case, at high frequencies, discontinuities in the shield make it act like a sieve. Poor RF performance of shielded cables due to shield imperfections arise, for example, in

1. EMP when large induced surface currents penetrate through openings, such as gaps, holes, imperfect seams, and so on, and cause damaging voltages inside metallic structures falsely assumed to be shields, such as airplane wings, ships, and cable armature.
2. Testing ceramic feed-through filters in text fixtures made such that the transmitter and receiver are hooked up to it by braided coaxial cables. They leak so much that filter measurements are falsified (the filter shows lower insertion loss because of by-passing via input/output coupling by the cables). About above 10 MHz, braided cables must be replaced by solid copper.
3. Improperly installed control systems, in particular where "well-shielded" cable bundles are virtually made unshielded if openings are left at the end of the shield. For instance, a HF-heated, automated plastic molding machine behaved erratically because the installer had twisted the shielding braids surrounding the control cables into a pigtail which he then screwed to "ground" (dc grounding versus HF grounding). Shield continuity was established by installing a backshell.

Based on calculations made by Kaden [2], Table 2.3 and Figure 2.3 set forth the key relations in nondimensionalized presentation.

Table 2.3 Transfer Impedance

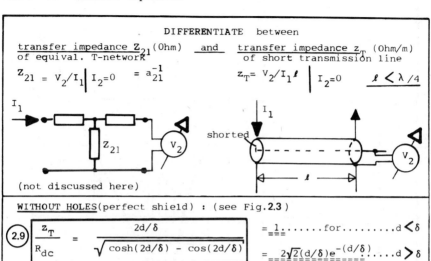

DIFFERENTIATE between
transfer impedance Z_{21} (Ohm) of equival. T-network **and** transfer impedance z_T (Ohm/m) of short transmission line
$Z_{21} = V_2/I_1 \big|_{I_2=0} = a_{21}^{-1}$ $z_T = V_2/I_1 \ell \big|_{I_2=0}$ $\ell < \lambda/4$

(not discussed here)

WITHOUT HOLES(perfect shield) : (see Fig. 2.3)

(2.9) $\quad \dfrac{z_T}{R_{dc}} = \dfrac{2d/\delta}{\sqrt{\cosh(2d/\delta) - \cos(2d/\delta)}}$

$= 1 \ldots\ldots \text{for} \ldots\ldots d < \delta$

$= 2\sqrt{2}(d/\delta)e^{-(d/\delta)} \ldots\ldots d > \delta$

d = thickness of shield δ = skin depth

WITH HOLES(imperfections) :

p = degree of perforation = $nr_o^2/2r_s$

According to Kaden [2]:

$z_T = (2/3\pi^2)j\omega\mu\, pr_o/r_s$
$R_{dc} = 1/2\pi r_s d\sigma$

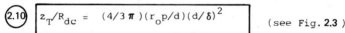

n = # of holes/length

$z_T/R_{dc} = (j2/3\pi)pr_o d\,\omega\mu\sigma$ and since $\delta^2 = 2/\omega\mu^x\sigma$ for $\mu_r = 1$

(2.10) $\quad z_T/R_{dc} = (4/3\pi)(r_o p/d)(d/\delta)^2$ (see Fig. 2.3)

IMPORTANT: For the same degree of perforation p, it is much worse to have one large hole than many small holes

FOR DETAILED THEORY, see:

Transferred Surface Currents in Braided Coaxial Cables, J.R. Martin, ELECTRONICS LETTERS, 7 Sep. 1972;

Shielding Effectiveness of Braided Wire Shields, E.F. Vance, IEEE TRANSACT., G-EMC, May 1975;

Application of Modal Analysis to Braided Shield Cables, K.S.H. Lee & C.E. Baum, IEEE TRANSACT. G-EMC Aug. 1975.

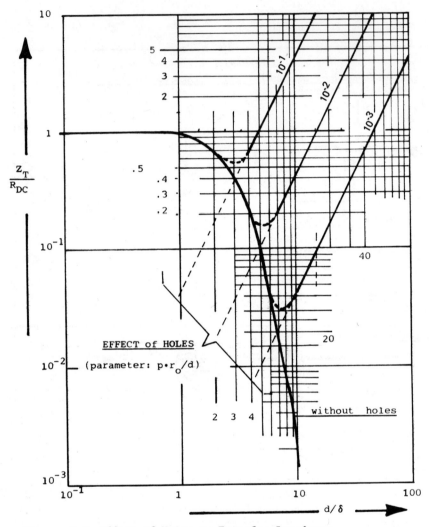

Figure 2.3 Effect of Holes on Transfer Impedance

Without holes or gaps (poor welds or riveted seams, in particular when vibrated), the coupling between outer cable currents into internal wires approaches zero for increasing frequency (Eq. 2.9, damping). But even small openings become more detrimental the higher the frequency. Table 2.3 includes references for more detailed theory. But for practical purposes, Table 2.3 sets forth all the information needed. But see also problem 99, Chapter 12.

Conductive Transfer 69

2.1.1.4 Modes of Propagation

For a large control system, the spreading (or branching or redundancy or whatever we call it) of transfer is made more difficult to analyze because of the modes of propagation possible. For radiative transfer there are different modes of polarization (and for our purposes we can disregard the various modes of long-distance propagation).

More important in this context is conductive transfer, where we must take into account a variety of modes, which we shall outline briefly:

1. *Directional modes.* Two-ports having unidirectional properties (energy or information flowing only in one direction) are very desirable for EMC work in that they restrict propagation. Three major classes of nonreciprocal devices can be identified.
    A. Amplifiers (active elements) acting as buffers, as, for instance, operational amplifiers.
    B. Isolators based on anisotropic material, as, for instance, microwave isolators.
    C. Many energy conversion devices are one-way devices, as, for instance, optoisolators (see later).
2. *Normal mode/common mode conversion in normal wiring.* In two (or more)-wire systems, the coupling is often such that both (or all) wires have current flowing in the same direction, with the return wire often being the ground. We discussed such common mode propagation in Table 1.8 and found that unbalance (different impedances in the wires) leads to common mode/normal mode reconversion.
3. *Waveguide propagation (not-TEM waves).* By either TE (or H) waves or TM (or E) waves, so named because only the electric field vector--or in the second case the magnetic field vector--is transverse to the direction of propagation, whereas longitudinal H or E components exist (in contrast to the ordinary TEM propagation--E and M transversal--characterizing low-frequency propagation in two-wire systems). TE or TM propagation

takes place in tubes or rods of material having dielectromagnetic properties different from those of the environment. The propagation is more perfect (no leakage) the more the ratio of the material properties approaches $0/\infty$ or $\infty/0$. Hence two types of waveguides based on internal reflections exist. Besides these modes, there are waves guided along the outside of single wires. Hence we have, altogether:

A. *Metallic waveguides*. Here waves are propagated inside metallic tubes of all kinds of cross sections. Figure 2.4 provides all the information needed. The waveguides are, without exception, *highpass* filters with hardly any attenuation above the cutoff frequency. Many modes are possible. The lowest mode, in a circular metallic tube, is the TE_{11} mode (describing a certain spatial field distribution, the details of which are of no concern here). For the TE_{11} mode, the cutoff frequency is 178 MHz for a 1-m-diameter tube and 17.8 GHz for a metal tube of 1 cm inside diameter. The highpass characteristic of waveguides is best observed when driving through a tunnel: One gets FM reception but no AM reception (stopband). The high attenuation of waveguides at low frequencies is of great practical significance: Holes in shields leak very much at high frequencies. But holes are often necessary (e.g., for ventilation). As Figure 2.4 indicates, changing the hole into a short tube

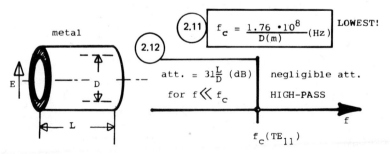

Figure 2.4 Lowest Cutoff Frequency of Tubular Waveguide

makes a great difference. A tube three times as long as its diameter reduces the leakage by 93 dB for low frequencies. If we need a large effective opening, we can nevertheless provide a very high cutoff frequency by using many small tubes in parallel (honeycomb filters).

B. *Dielectric waveguides.* These are used primarily to propagate extremely high frequencies, or infrared and visible light. For short runs of fiber cables, the attenuation is negligible. The introduction of fiber optics fed by light-emitting diodes (LEDs) for control systems and fed by lasers for communications systems eliminates interference coupling and is, therefore, a control tool of utmost importance for EMC.

However, it is a serious mistake to simply drill a hole into a metal shield to pass the light pipe through without providing metallic waveguides (see before) to prevent leakage of very high frequencies.

C. *Surface guided waves.* For completeness, we mention here in passing guiding wires, that is, single wires having a rough or dielectric surface. Excitation is possible only by feeding with horn antennas.

We switch now to some time-domain relations that are very important in conducted transfer.

2.1.2 Time Domain

We leave exacting exercises in Laplace transform to college professors. Here we are only interested in the essence of time/frequency relations as we need them to make reasonable estimates of the OOM of transients and their parameters.

2.1.2.1 Rise Time and Bandwidth

For signal lines we are justified to work with the approximations set forth in Table 2.4 (tacitly assuming essentially matched conditions and linear phase). Equations 2.13 and 2.14 approximate

Table 2.4 Rise Time and Bandwidth (Sine Integral Approach)

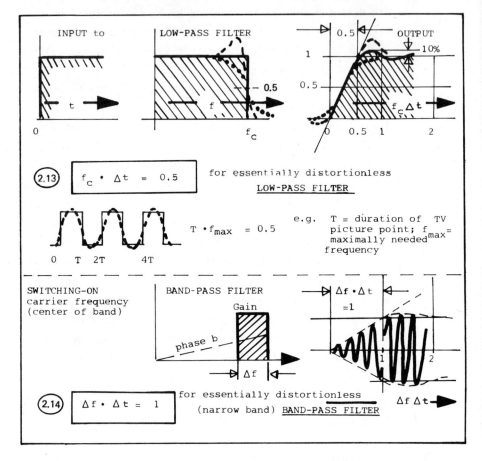

the transformation from time domain to frequency domain and reverse. The factor 2 difference in the two equations lies in the fact that for Equation 2.13, the dc component is included; for Equation 2.14, it is not. But in both cases, the reciprocity of determinacy of rise time and frequency is the key relationship of great practical significance. The smaller the bandwidth, the more indefinite (sloping) is the response to a step function.

We deviate here from the conventional definition of rise time (defining the slope between 10 and 90% of amplitude and figuring

$f_0 \Delta t = 0.35$). Rather, we take the maximum slope of the sine integral, which is the response to an ideal low-pass filter.

Let us keep this inverse relationship of f_c and Δt in mind, as this fact gives us an excellent tool to reduce rise times.

Example 2.6 When is the full amplitude reached if a step function is switched into a matched load via a low-pass filter having f_c = 50 Hz? Δt = 10 ms.

2.1.2.2 Delay and Rise Time on Transmission Lines

In Chapter 1 we saw that the response of a network to an infinitely steep step function is immediate only (zero rise time) if the network contains only resistors, but that the presence of reactances always results in finite rise times.

If, in contrast, we apply a step function to a lossless transmission line, the response at the end of the line will also have zero rise time, although the step (or pulse) is delayed by the time needed to travel along the line with the speed

$$v = \frac{3 \times 10^8 \text{ m/s}}{\sqrt{u_r \varepsilon_r}} \quad (2.15)$$

This is not in contradiction to Section 2.1.2.1, inasmuch as a nonlossy transmission line is an all-pass network having $f_c = \infty$. However, if the line is lossy (distributed resistance elements) an infinitely steep wave front will be flattened--in addition to the delay--as f_c is no longer infinity. As the wave moves along the line, it needs finite time to penetrate the conductor. In the beginning, the propagating wave is penetrated less in the conductor (fewer losses involved) than later. Hence the wave front is initially rather steep and becomes more and more flat the more the field penetrates into the conductor.

The composite nomograph in Figure 2.5 permits calculation of the wave front formation just described. Here we look only at the effect of eddy current losses in a two-wire line (dielectric losses are neglected). The Laplace transform yields the response in the

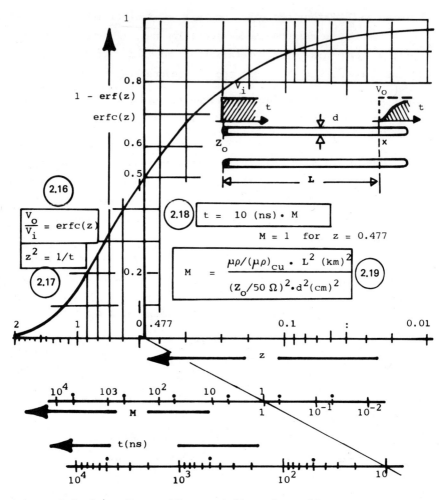

Figure 2.5 Steep Fronts Flattened Along Lossy Lines

form of the Gaussian error function (Eq. 2.16) of the variable z, which is proportional to $t^{-0.5}$ (Eq. 2.17). For z = 0.477, 1 - erf(z) = 0.5. For this half-amplitude point, t = 10 ns for M = 1 as defined by Equation 2.19. Equation 2.19 displays some rather unexpected quadratic relations: A wave front is made *four times* as long [but still maintaining the 1 - erf(z) configuration] if we *double* the length of the line, or *half* the characteristic impedance of the wire diameter. In contrast, material constants enter linearly. M = 1 for a copper line having the following parameters: for example, Z_c = 50 ohm, $\mu_r \rho$ for copper, L = 1 km, and wire diameter = 1 cm. Hence for such a line, by definition, 50% of the pulse front is reached in 10 ns, but 90% in 300 ns (use the nomograph).

Example 2.7 For the line just discussed, we replace the copper by iron, the $\mu_r \rho$ of which is 500 times larger. The nomograph yields 5000 ns for the half-amplitude point. For more information, see Ref. [3].

2.1.2.3 Nonlinear Effects in Op Amps

In Section 2.1.2.1, we sketched the basic reciprocal relationship between bandwidth and rise time, tacitly assuming linearity. Many systems incorporate amplifiers for which Equation 2.13 holds only for the small amplitudes for which they are designed. Voltage spikes can easily invalidate this condition: Unless the amplifier is destroyed, it may saturate or ring (phase margin changed). While clipping, the pulse can eliminate these nonlinear responses; it cannot nullify the error that can still be caused by the clipped pulse.

There is another nonlinearity in op amps; the slew rate. In general, one tries to make it as high as possible. But for EMC purposes, we exploit this nonlinearity by expressly reducing the slew rate to get rid of sharp voltage spikes occurring within the signal envelope (see Section 5.3). For details of these nonlinearities occurring in op amps, see Ref. [4].

2.2 RADIATIVE TRANSFER

The confining properties of shielding and absorbing are often misunderstood or missimplified in that the large ranges of the key criteria FATTMESS are not taken into account (frequency, amplitude, time, temperature, mode, and most important, size in terms of wavelength). We shall try to delineate various lower- and upper-limit asymptotic cases which we can handle with a modest amount of mathematics, with the emphasis on gaining an understanding on the governing mechanisms. In most cases, the more complex conditions between lower- and upper-limit approximations can be sketched in by hand with sufficient practical accuracy.

2.2.1 Frequency Domain

2.2.1.1 Unimpeded Transfer

Table 2.5 compiles a few very useful relations. They permit a quick estimation of the amplitudes we may expect for unhindered transfer. Communications engineers are quite familiar with the content of Table 2.5, and engineers concerned with good EMC of control systems should familiarize themselves with them:

An approximation of the field strength as a function of transmitter power and distance from the transmitter.

The relation between electric and magnetic field strength (free space).

The radiation impedance of wire antennas. In particular, Example 2.8 illustrates the poor radiation efficiency of short wires at low frequencies.

2.2.1.2 Reduction of Transfer

Before doing some rather instructive calculations, let us list a few rules of thumb pertaining to the attenuation caused by buildings:

Normal building erected without any EMC precaution dampens about 10 dB.

Buildings with a double layer of rebars (carefully welded) may attenuate by about 25 to 35 dB.

While we shall consider more complex problems of shielding in Chapter 7, here we review the propagation aspects in more detail. Yet, admittedly, the division of material between Chapters 2 and 7 may look somewhat arbitrary--it is like the chicken-and-the-egg problem. At any rate, in either case, we approximate our shielding or absorbing boxes by spheres in order to take the important size/wavelength parameter into account.

Radiative Transfer

Table 2.5 Key Radiation Relations

(a) **RECEIVED FIELD STRENGTH**: E received from sender of power P and located at distance d:

$$E = 3 \cdot 10^5 \sqrt{P}/d \quad (2.20)$$ where $E(\mu V/m), P(kW), d(km)$

or, in dB/uV/m, as many instruments are calibrated:

$$E(dB/\mu V/m) = 20 \log E$$

(b) **FREE SPACE IMPEDANCE**: Z_c^* is the ratio of electric fieldstrength to magnetic fieldstrength for plane waves

$$Z_c^* = E/H = 120\pi = 377 \text{ ohm} \quad (2.21)$$

(c) **RADIATION IMPEDANCE** of a vertical wire antenna is given by a spiral diagram

+jX, thin wire, very thick wire, $\lambda/4$, R_s, $\lambda/2$ 200 ohm, $\lambda/2$ 2000 ohm, -jX, 37 ohm

hence for good broad-band efficiency use very thick antenna (circular wire groups)

(see discussion of η, before)

Most inefficient: very short antenna: $h \ll \lambda$

for which radiation resistance $$R_s = 1600 \ (h/\lambda)^2 \text{ ohm} \quad (2.22)$$

EXAMPLE 2.8: EFFICIENCY OF VERY SHORT ANTENNA: (short in terms of λ) An = 20 m antenna is driven by a 15 kHz sender (λ = 2000 m). The antenna tuning coil must have a reactance of +j1600 ohm. With great effort, using broad Litzwire strips, an excellent Q of 1000 can be achieved. That means the loss resistance of the coil is 1.6 ohm. Hence, disregarding the ground resistance, the efficiency is 0.0016/1.6 = 0.1%. This example shows that we do not have to be much concerned about low-frequency radiation (far field) and also explains why project SEAFARER needs enormous antennas

---

Waves Coming from Outside (Quasistatic Fields)

Table 2.6 refers to a pair of complementary equations for quasi-electrostatic and quasistatic magnetic fields. The table is self-explanatory and contains Example 2.9, electric-field-only shielding.

Table 2.6 Dielectromagnetic Spheres at Low Frequencies

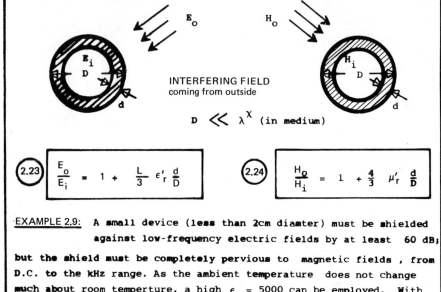

EXAMPLE 2.9: A small device (less than 2cm diameter) must be shielded against low-frequency electric fields by at least 60 dB; but the shield must be completely pervious to magnetic fields, from D.C. to the kHz range. As the ambient temperature does not change much about room tempertaure, a high ϵ = 5000 can be employed. With E_o/E_i = 1000 = 1 + 1.33x 5000xd/5cm , the wall of a 5 cm sphere must be at least 0.75 cm.

Equations 2.23 and 2.24 also explain why for dc or very low frequencies, metallic shields (high ϵ_r, Eq. 2.3) are highly effective for electric fields, but not for magnetic fields, unless the metal is magnetic or eddy currents are generated by sufficiently high frequencies.

Waves Generated Inside a Sphere (High Frequencies)

By operating with spherical coordinates, we can extend the transmission line concept (Table 2.1) for point sources in spheres having all kinds of boundary conditions. Here the concept of radial impedance facilitates considerations. The key points are set forth in Table 2.7, which contains two brief examples to be followed by a detailed discussion of a neoteric shielded room. The reader should make an effort to grasp the essence of these examples,

Table 2.7 The Concept of Radial Impedance

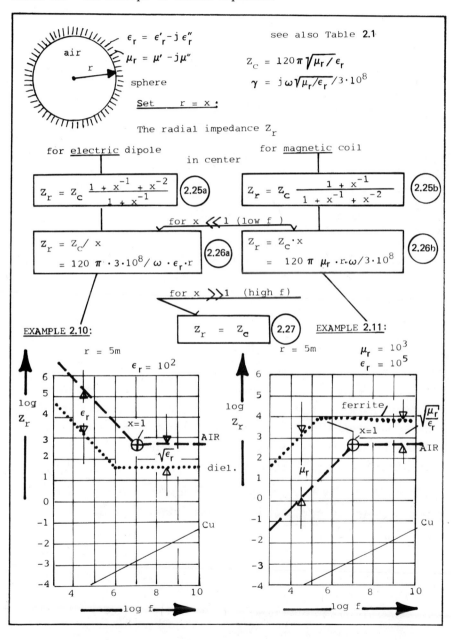

as they clearly demonstrate the importance of dimensions and material "constants," in particular complex μ_r and ε_r.

For "extreme" propagation measures $\gamma r = x$, we find very distinct asymptotic behavior, depending whether $x \gg 1$ or $x \ll 1$. For both conditions, the original equations are very much simplified--to Equations 2.26a and 2.26b.

For a spherical interface of radius r, we find the ratio of interface impedances (medium/air) to be

$$\frac{Z_r}{Z_r^*} = \frac{377(\mu_r/\varepsilon_r)^{0.5}}{377} = \left(\frac{\mu_r}{\varepsilon_r}\right)^{0.5} \qquad (2.28)$$

$$\underline{x \gg 1}$$
Electrical and magnetic case

For electric dipole: $\dfrac{Z_r}{Z_r^*} = \sqrt{\dfrac{1}{\varepsilon_r}}$ \hfill (2.29a)

$$x \ll 1$$

For magnetic coil: $\dfrac{Z_r}{Z_r^*} = \sqrt{\mu_r}$ \hfill (2.29b)

See Examples 2.10 and 2.11, Table 2.7.

We now apply the concept of radial impedance to a conducting medium characterized by a dominant $\varepsilon_r'' > \varepsilon_r'$, at least for lower frequencies (see Eq. 2.3 of Table 2.1). We use the conductive medium as an absorber, het we must make sure that the energy to be absorbed really penetrates the medium and is not reflected from its surface--at least at higher, critical frequencies.

Example 2.12 (The Hole Underground) Roughly, design an unconventional "shielded room" to meet the following:

1. *Objectives*
   A. Sphere of 5 m radius to accommodate vehicles (to be EMC tested, in particular for susceptibility of microprocessors installed in them).
   B. Shielding effectiveness 80 dB, 1 MHz to 1 GHz.
   C. Negligible standing waves.

D. In the interest of low cost, use soil as absorber (10 ohm-m $< \rho <$ 1000 ohm-m; wet to dry) ($5 < \varepsilon_r' < 10$).
   E. No need for costly welding and expensive doors.
   F. Operation in all weather conditions (freezing!).
2. *Statement of key problems and alternatives.* For sufficiently low frequencies ($2r < \lambda$), we have a nearly constant field distribution, but for higher frequencies ($2r > \lambda$), pronounced standing waves may introduce pronounced measuring errors. Conventional microwave absorbers must be ruled out because of fire hazards, under high power excitation, and because of high price. We prefer soil because it is amply available and is a good absorber. But it has two limitations which must be circumvented:
   A. Frost will drastically reduce soil conductivity, hence also greatly reduce absorption. The way out: Go below the frost line, apply heat, or put under the house.
   B. Front surface reflection must be minimized such that transmitter energy is guided into the absorbing soil to avoid standing waves. We are interested in resonances of the vehicle standing on ground, but want to avoid resonances of the enclosure.

   Two remedies are at our disposal to overcome the reflection problem: First, we can work in the time domain (very short high-energy pulses feeding a more or less matched transmission line with the vehicle in its fringing field). Not only would the front surface reflection be less critical, but the instrumentation would be simpler. Moreover, lightning and EMP can be simulated. Second, we can work with a nonburnable transition layer, which we discuss after performing some calculations.
3. *Asymptotic calculations.* Figure 2.6 presents the results of calculations for which the foregoing discussions have prepared us. The radial impedance of the 5 m air sphere is a singly broken line (corresponding to Example 2.10). For the r = 5 m soil sphere, the radial impedance is a double broken line: For very high frequencies, ε_r' is dominant; thus we get a horizontal

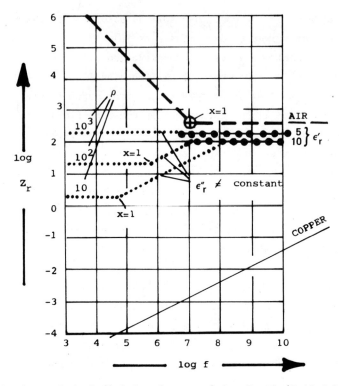

Figure 2.6 Radial Impedances of 4 = 5m Air/Soil Spheres

line, as for air. For lower frequencies, ϵ_r'' dominates and we have to employ Equation 2.3, 2.28, or 2.29a, respectively, depending on $1 < x < 1$. We calculate $f_1 = \rho$ (ohm-m) $\cdot$ 5100 for which $x = 1$. This determines the second kink in our lines for the radial soil impedance. Above $x = 1$, $Z_r = 0.0028\sqrt{f\rho}$. It increases with the same square-root-of-f slope as copper does. On the other hand, at very low frequencies ($x < 1$), Equation 2.29a holds and $Z_r = (2/10\pi)\rho$, resulting again, in Figure 2.6, in a horizontal line.

4. *Transition layer.* Figure 2.6 shows that, for the critical higher frequencies, front surface reflection is already quite small. In fact, in actual operation under high-power excitation (200 V/m is a realistic number), the innermost layer of

Radiative Transfer

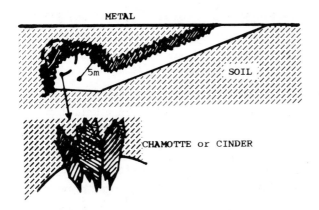

Figure 2.7 Double-Edged Transition Layer

soil will be so heated that it becomes rather dry and therefore even less reflective.

But since (the test object) and the energy sources are not the point sources upon which our calculations are premised, we have waves of oblique incidence. That means higher reflection.

Thus, to suck the waves into the soil, we employ the interface layer depicted in Figure 2.7. The following advantages result from the low-dielectric-constant cinder or chamotte:

A. Structural constraints of soil (ceiling).

B. Double cones; gradual diffusing transition both into air space and soil space.

5. *Absorption.* If one relies on soil absorption alone, the thickness of the soil mass around the air bubble must be at least 10 skin depths: $10N = 87$ dB. But for the top, something like one δ is sufficient if the top is covered with continuous metal (which we need anyhow to mount common mode filters). As the soil is then traversed twice, the reflected wave is reduced by 17.4 dB, hence is negligible for the generation of strong standing waves.

To get an idea of the minimum thickness, we calculate as follows: At 1 MHz, Cu has a skin depth of 0.07 mm (Chapter 7). Since Cu has a resistivity of $\rho_c = 1.7 \cdot 10^{-8}$ ohm-m, soil at the

at the same frequency has $\delta_s = 0.07$ mm $\sqrt{\rho_s/\rho_c}$, that is, 1.7 m ($\rho_s = 10$ ohm-m), 5.4 m (100 ohm-m), and 17 m (1000 ohm-m).
There is still another advantage of our novel earthen "shielded" room. We have no critical metallic joints (gaps, slots). In fact, we can now subject our test objects without impunity to heavy mechanical vibrations or shocks which, in a welded or riveted metallic shielded room, could periodically open and close leaking gaps.

Figure 2.7 also indicates a lossy waveguide entrance, eliminating the need for expensive, tightly fitting doors, provided that the access tunnel is sufficiently long and has a good transition layer.

2.2.2 Time Domain

Even for unimpeded transfer, we have to take into account time delay, due to finite propagation velocity (Ea. 2.15). In most cases, the delay is insignificantly small (e.g., in air a wave front travels 30 m in 1 ns). By impeded transfer (shields), a steep wave front can be considerably flattened, provided that the shield is perfect (no openings).

Inasmuch as we are presently concerned with EMI analysis, we look here only at the very idealized case of an infinitely large plane shield, in order to get a feeling for the dB/dt reduction imparted by shields that is so important in regard to EMC. And we relegate approximations closer to reality (nonlinearity and size effects) to the control part (Chapter 7).

Based on the detailed description given in Ref. [5], Table 2.8 provides the information necessary to calculate the dB/dt reduction for the plane linear case.

Example 2.13 (set forth in Table 2.8) refers to a pulse (zero rise time) traveling along a wire and introducing a rising voltage into a suitably placed loop. We juxtapose the dB/dt and the induced voltage for two sets of conditions:

Table 2.8 dB/dt Reduction Imparted by Shields

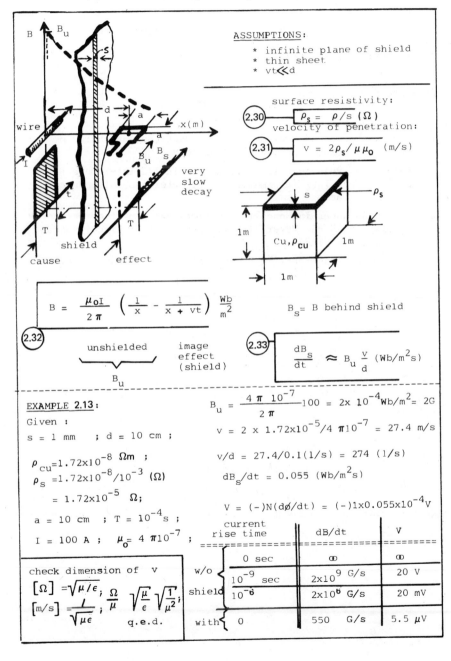

1. Without a shield for zero, 10^{-9}, and 10^{-6} rise time of the current pulse
2. With a 1 mm copper shield for zero rise time of the pulse

The reduction is striking.

REFERENCES

1. IEEE Standard 518-1982 (Noise Guide).
2. Wirbelstroeme und Schirmung in der Nachrichtentechnik, H. Kaden, 2nd ed., Springer-Verlag, Berlin, 1959.
3. Nanosecond Pulse Handling Techniques in I/C Interconnections, AN-270, Technical Information Note, Motorola Semiconductor Products, Inc.
4. Recent Advances in Monolithic Operational Amplifier Design, P. R. Gray and R. G. Meyer, IEEE Transactions of Circuits and Systems, May, 1974.
5. Electrical Transients in Power Systems, A. Greenwood, Wiley-Interscience, New York, 1971.

3 Receptors
An Unsuspected Multitude

We are concerned here primarily with incidental or quasi-incidental receptors and very little with susceptibility as it pertains to narrow-band radar or communication receivers. The latter are well treated in military standards and conventional EMC books. Rather, our concern is about *any* potential receptor, intentional or not, that may cause malfunction, failure, or injury, directly or indirectly. We want to protect machines and human beings; thus we have to analyze the multifaceted vulnerability to a great variety of sources, imparting interference, degradation, or failure. We distinguish basically between nonliving and living receptors. In each case we must know cause and effect. The cause is often bi- or terconditional, often with one condition being dominant. Causes and effects are so multifaceted that, for completeness, we would have to classify them according to quite a number of different principles into unavoidably overlapping categories. And still we would have to fill the entire book. Many specific numerical values of susceptibility limits--though often incomplete--are available in manufacturers' data sheets and application notes. The EMC engineer should keep a continually updated file of such important information.

What we intend to do in this course, then, is not to compile manufacturers' data; rather, our goal is to provide the necessary understanding such that we can view these data critically for completeness and relevance. Moreover, there are many receptor mechanisms, quite a number of them unsuscepted, which are described extensively in the literature. To get a practical handle on this multifarious situation, we shall proceed as follows:

1. *Technical (nonliving) systems.* In order not to overlook important receptors we will view briefly three typical (often generic) examples for each of the eight FATTMESS attributes. Emphasis is on principles and perspectives such that the reader learns to realize the great variety of receptors and parameters to look for.
2. *Living receptors.* Here the situation is quite different. Although electrical hazard limits are well established, other, so-called "side" effects are still highly controversial. Here we must distinguish between statistically established facts and the conjecture of receptors for which the underlying mechanisms are often still quite obscure and for which the effects are often not easily reproducible.

3.1 TECHNICAL RECEPTORS (FATTMESS CATEGORIZATION)

The conventional approach is to classify by degree of the effect, such as insignificant temporary disturbance, serious disturbance, and catastrophic failure. Another classification is based on the duration of the cause or transient. Others are classified by the type of operating mode of the device, such as analog or digital. Here our basis for classification is a set of critical FATTMESS attributes. Our comments will be very brief or more elaborate, depending on the (admittedly subjective) assumption (of the author) about what he found to be generally known or not known and whether or not the data are readily available.

3.1.1 Frequency (F)

1. *Logic decision elements affected by high frequency.* Logic elements are biconditionally (frequency, amplitude) affected: for instance, HTL, any signal over 10 V peak to peak to 3 MHz; TTL, any signal above 3 V peak to peak up to 30 MHz; ECL, any signal 2 V peak to peak up to 150 MHz.
2. *Oscillators may be entrained, fully or partly.* Self-excited oscillators of the frequency f_0 are, by their very nature, unstable entities, the frequency of which can be fully or partly entrained by the external frequency f_1, if $m \times f_1 - n \times f_0 = \Delta f$ is about zero, where m and n are integers. The effect is most pronounced if m = n = 1, less if m or n = 1 and the other multiplier is not far from 1. The effect is hardly noticeable if m and n are much larger than 1. Figure 3.1 explains this phenomenon of synchronization: Nonlinearity is inherent in any oscillator. Otherwise, its amplitude would be infinity. Thus, unavoidably, the frequencies f_0 and f_1, and/or their harmonics, will form a modulation product M. Its frequency may be close to, but not completely identical with, the original oscillator frequency f_0. Hence the vector M of the modulation product is initially free to take any position with respect to the original unaffected vector S of the frequency f_0. The vector diagrams (a), (b), and (c) of Figure 3.1 show how differently M can be trapped. In Figure 3.1(a), the Δf is so small that $\tan \beta = 2\Delta f Q$ is smaller than M/S (where Q is the Q of the tank circuit). M is caught in the position given by $\tan \beta = 2\Delta f \times Q/f_0$. In (b), we encounter the limiting case of synchronization, and in (c), $\tan \beta = 2\Delta f Q/f_0$ is larger than M/S--only slightly, though. We lose synchronization. A peculiar phenomenon occurs. Now the actual beat frequency Δf_b is smaller than the Δf that would exist if f_0 and f_1 would not interact. And the beat frequency shows a characteristic sawtooth-like form. M, being a quasifree agent, rotates now with a momentary difference frequency that is close to zero (flat part of the beat) if the

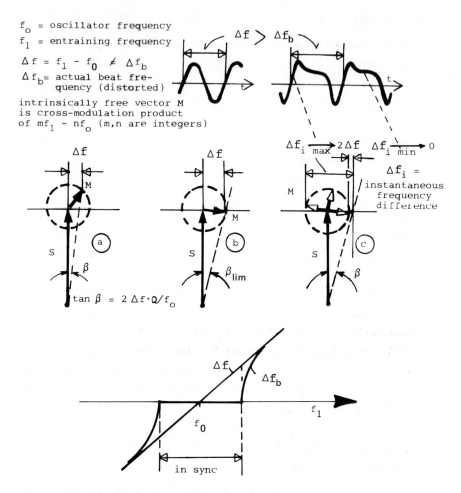

Figure 3.1 The Mechanism of Synchronization

vector end point is close to the dashed line of β and nearly twice the undisturbed Δf for the opposite position. In fact, at the right-hand side, M moves very slowly, as it is nearly trapped into this position: A bit smaller than Δf or a bit larger than M would catch M in the limiting position depicted in vector diagram (b). Thus, to prevent frequency entrainment, including subharmonics, buffer stages, filtering, and shielding must be applied.

3. *Penetration is frequency dependent.* The higher the frequency and the conductivity, the smaller is the depth of penetration of current (skin depth, Chapter 7). Current flow and initial heat effect will, therefore, be localized at the surface for high frequencies and good conductivity, whereas at low frequencies the current finds less resistance and flows nearly evenly over the whole cross section; also, the heat effect is rather evenly distributed. Thus a person may be shocked or killed when subjected to 120 V ac but not for microwave frequencies (for which other effects occur, as we will show shortly).

3.1.2 Amplitude (A)

1. *Dielectric breakdown starts wherever E is high.* There is no magnetic breakdown, although mechanical deformation of conductors can occur due to very high magnetic fields. But electric breakdown of dielectrics originating in areas of high electric field strength E is a common cause of EMC problems. In general, the withstand voltage decreases with increasing temperature. See also Section 1.1.2 for field concentrations and Section 3.1.3 for time effects, and physics books.
2. *Amplifiers may not perform as planned.* Instrumentation amplifiers may saturate or ring strongly if subjected to voltages above rating, as we discussed in Chapter 2. This may render them useless or interfering.
3. *Other nonlinear effects arise in polarizable materials.* Saturation may negate the effectiveness of filters, in particular for higher temperatures in the case of magnetic saturation and for both lower and higher temperatures for high-ε ceramic dielectrics. See also Section 1.3.2.3.

3.1.3 Time (T)

1. *Dielectric breakdown is time dependent.* The breakdown of a dielectric is not an immediate response to an applied voltage. Rather, the formation of the breakdown requires time for carrier

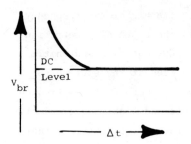

Figure 3.2 Breakdown and Pulse Width

formation and for avalanching. This depends on the voltage supplied, on the surface state of the dielectric, on its impurity content, and on its temperature. Figure 3.2 shows the general behavior of dielectric breakdown as a function of time (breakdown voltage versus pulse width). Dielectrics withstand much better short pulses in that their short duration does not permit full formation of the breakdown process. The location of the knee in Figure 3.2 varies from milliseconds to nanoseconds, depending on whether we have ionizing liquids or fast-switching spark gaps.

2. *Logic circuits have switching delay.* Figure 3.3 pertains to representative switching delays of logic families and is surprisingly similar to Figure 3.2. The horizontal part of the curve characterizes the static noise immunity; the descending

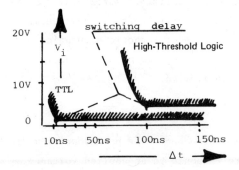

Figure 3.3 Switching Voltage and Pulse Width

curve characterizes the dynamic noise immunity. If the pulse width Δt of an interfering pulse is smaller than the switching delay of the system, the permissible amplitude may be considerably larger (energy) than in the static case. In Chapter 5 we show how we can intentionally increase the delay of latches in order to make them nonresponsive to short pulses.

3. *Corrosion and deterioration show up later.* Delays of weeks, months, or years are involved in the slow deterioration of dielectrics (precorona; see Section 1.1.2) or conductors (corrosion due to the formation of galvanic elements by joining unequal metals in an unavoidably polluted atmosphere). In the latter case, grounding or bonding connectors, though initially solidly soldered or welded, may disappear. Havoc may follow. Even such simple things as resistors deteriorate under continuous pulse conditions. The most stable resistors are hot-molded carbon composition resistors, which degrade only slightly under long exposure to pulse operation, with the average power not exceeding the rating of the resistors. On the other hand, film resistors, under such conditions, degrade continually until they fail (open circuit), thin films quite a bit faster than thick films (cermet).

3.1.4 Temperature (T)

1. *Secondary breakdown of semiconductors does not need much power.* Power semiconductors may switch current very fast, let's say on the order of 10^9 A/s. Then the junction can develop hot spots, causing failure. But more relevant is the secondary breakdown of semiconductor junctions caused by reverse pulse voltages [1]. Its failure susceptibility lies about 1 decade below the power rating of the semiconductor and is caused by a Zener or an avalanching effect. Reference [1] summarizes Wunsch's thermal model. We simplify it here and work only with the two asymptotic cases instead of his original three cases. If we

plot, logarithmically, the threshold power versus pulse with Δt, we can distinguish two asymptotic ranges. Below $\Delta t = 1$ µs, the power declines with $1/\Delta t$, as the heat cannot flow sufficiently fast that the limiting parameter is the constant energy. Above about 10 µs of Δt, the time is sufficient for more or less even heat distribution such that we can approximate larger pulse widths with a line of constant power versus Δt. The transition range lies, then, between 1 and 10 µs and is rather broad. Upon request, many semiconductor manufacturers will provide pertinent information on pulse amplitude or energy versus Δt. Table 3.1 should help in conversion and checking dimensions.
2. *Frost nullifies grounding.* As we saw in Chapter 2, frost may reduce the conductivity of soil by many decades. This is particularly significant in the arctic; but even in the northern United States, in cold winters, the frost line may be 2 m deep.
3. *Temperature "softens" many a material.* Elevated temperatures not only predispose isolators to have lower withstand voltage (already discussed) but also may reduce desirable material properties. For instance, thermoplastic materials lose their mechanical strength (causing displacement and failure). Or heat may drastically reduce the saturation density of ferrites (the Curie point of ferrites is on the order of 300°C; of steel it is 800°C)--thus reducing the effectiveness of high-frequency chokes (ferrite beads and baluns) or ferrite cores used for transformers in switching power supplies.

3.1.5 Modes (M)

1. *Polarity can be critical.* Earlier we saw the dangers of reverse voltages. As the polarity of the pulse is of such importance for semiconductors, we list it as a special criterion. For some transistors, a 10 V reverse polarity spike can be destructive.
2. *RF changes transistor characteristics* (see also Sec. 3.1.1). Reference [2] is the most recent comprehensive investigation of high-frequency effects on linear and digital integrated circuits.

Table 3.1 MKS(r) System

Symbol	Designation	Dimension m	l	t	q	Unit	Important Relations
m	Mass	1				kg	
l	Length		1			m	
t	Time			1		s	
q	Charge				1	C	$W/V = It = Q = CV$
f	Frequency			-1		Hz	$W/V = It = Q = CV$
v	Velocity		1	-1		m/s	$f = v/\lambda$
a	Acceleration		1	-2		m/s²	$W = mv^2/2$
	Impulse	1	1	-1			$F/m = a$
F	Force	1	1	-2		kg	$Ft = A/1$
A	Action	1	2	-1		Js	$W/1 = QE = F$
W	Work (energy)	1	2	-2		J	$= 1.53 \times 10^{33}$ quanta
P	Power	1	2	-3		W	$F1 = Pt = QV = W$
I	Current			-1	1	A	$VI = W/t = P$
V	Voltage	1	2	-2	-1	V	$Q/t = I$
E	Electric field	1	1	-2	-1	V/m	$W/Q = IR = P/I = V$
H	Magnetic field		-1	-1	1	A/m	$V/I = F/Q = E$
Z	Impedance	1	2	-1	-2	ohm	$I/1 = H$
μ	Permeability	1	1		-2	$4\pi \times 10^{-7}$ (H/m)	$V/I = Z$
ε	Dielectric constant	-1	-3	2	2	$(36\pi \times 10^9)^{-1}$ (F/m)	
L	Inductance	1	2		-2	H	$2\pi fL = Z$
C	Capacitance	-1	-2	2	2	F	$2\pi fC = Z^{-1}$

Nondestructive reversible interference is essentially caused by rectification in the various semiconductor junctions and may cause dc offset of the output voltage. This effect tapers off with higher frequency (capacitance). Degradation and failure appear at higher power levels, with linear circuits being generally less susceptible.

3. *Latching of C-MOS devices is costly.* C-MOS devices may latch up, that is, may form an SCR which may mean a permanent switched-on position or destruction. Latch-up occurs if the analog input signal is larger, even if only for a very short time, than the power supply voltage. See, for instance, the detailed application notes by Intersil. They cite the following event: unplugging and replugging-in an electronic game machine caused a payout, the cause of which was fully unsuspected.

3.1.6 Energy (E) and Power

1. *Survey of energy effects* (see also Sec. 3.1.4.1). Ranging over 17 decades of energy, Figure 3.4 gives a survey of the minimum impulse energy causing (a) erroneous outputs and (b) destruction of typical devices. In specific cases, the manufacturer's data should be consulted, but for an estimate, Figure 3.4 provides an order of magnitude indication, if short, let's say 1 μs, pulses are involved. Destruction is caused by localized heat. Keep in mind that $1 V \times 1 A \times 1 s = 4.18$ cal.

2. *Power counts for long or repetitive pulses.* For longer pulses (larger than 0.1 ms) and repetitive pulses, power counts if the thermal equilibrium is formed before destruction sets in (see Ref. [1], p. 00).

3. *Flammability is consequential.* Ignition of explosives and combustible materials and fire of flammable materials may not be caused only by arcing or sparking but also by a slow localized heat buildup due to electric field concentrations. Many materials burn or give off noxious gases. Some materials can be made flame retardant or flameproof.

Technical Receptors (FATTMESS Categorization)

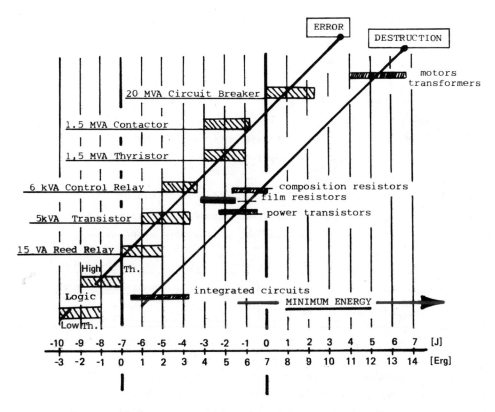

Figure 3.4 Errors or Destruction Caused by Pulses

3.1.7 Size (S)

1. *Resonances must be taken into account.* Quite often, parts or components are interfered with or damaged by intermediate receptors, in particular, those which resonate. Then, in the case of transients, amplitude multiplication by 2 (for high-Q switching) and by Q for continuous excitation may occur (see Chapter 1). Such amplitude-enhancing resonators may be transmission lines, automobile bodies, airplane wings, enclosures, and so on. For instance, an airplane wing may act as a $\lambda/2$ antenna, which in the center of the airplane, if EMP excited, may carry 20,000 A peak slowly decaying at a frequency

3×10^{10} cm/s divided by $2 \times \lambda/2$. Such electromagnetic resonances are normally well above 1 MHz. Q reduction, by increasing R or decreasing Z_0, is often feasible and quite effective. Mechanical resonance frequencies of enclosures are in the range hertz to kilohertz. Mechanical vibrations may periodically open and close gaps of riveted or poorly welded structures. These vibrating gaps intermittently admit energy to inside equipment. Hence EMC testing should be done under operational mechanical vibration and shock.

2. *Dirty insulators arc over.* Surfaces of insulators, unless properly corrugated, may accumulate dirt and grime, which, in the presence of high humidity, causes corona and arcing.
3. *Beam leads may burn away.* Beam leads of integrated circuits, because of their very small cross section, may burn away (see Ref. [1], p. 94). Beam leads, then, act like fuses--but they are very hard to replace.

3.1.8 Statistics (S)

1. *Components have limited life.* The unavoidable variability of manufacturing processes causes a bell-shaped Gaussian normal curve distribution of specific properties if plotted versus number of occurrences or a straight line in cumulative probability paper, as we discussed in conjunction in Figure 1.10. If, instead of a straight line, a broken line occurs, two dominant failure mechanisms can be suspected. Components for which quality control tests show such breaks in the distribution line are not acceptable for high-reliability systems. Reliability is a probability of acceptable performance of a component or system for specified conditions for a specified time. Plots of the average failure rate versus time are typically as follows:
    A. It is beginning as a quick decline of the failure rate (burn-in for semiconductors or debugging of the whole system).

B. This is followed by a much larger period of normal operation, where the failure rate is essentially constant. For semiconductor devices, this normal operating time is much larger than for vacuum tubes or relays.
C. At the end comes the wear-out portion, where the failure rate increases again.

The reciprocal of the failure rate is the mean time between failure (MTBF). The limited life of system components--though not strictly EMC--must be taken into account in EMC planning, as these effects are often the same as those of EMC proper or may even be caused by unsuspected EMI.

2. *Analog versus digital criteria.* The conventional distinction between "signal-to-noise ratio" in analog circuits and "probability of error" in digital circuits is often quite appropriate. For details, see EMC handbooks and standards. But such rather simple criteria may be rather inappropriate and even very costly in very noisy systems or environments, where the selection of different signal detection and handling methods may be more cost-effective (see below and Chapters 5 and 7).

3. *Signals may only seem to be buried inextractably in noise.* We conclude our survey on nonliving receptors with some teachings of information theory: If we have noise affecting our intentional receptors, such a receiver will generally detect a specific kind of signal for which it is optimized or tuned, as, for instance in the simplest case, a radio receiver selecting only a narrow, designated frequency band. We know that FM is much more noise-immune than AM. But we can go much further than that. By identifying differentials of signal and noise in the frequency, time, and statistical domains and by proper signal and detection design, we can often operate in quite noisy environments without disturbances. We outline such approaches later, only hinting here at the use of blanking and timing (time domain) or correlation techniques (statistical domain).

3.2 LIVING RECEPTORS

We divide this important topic into two parts:

1. The direct, coarse effect of power feed line currents or very short high voltage pulses on the human body--with the effects ranging from a mere tingling to death and charcoaling
2. The much more subtle "side effects" spanning the whole frequency spectrum and ranging from discomforting very low frequency alpha-wave entrainment to microwaves causing cataracts in the eyes

In item 1, we indicate various danger limits defined at least in their order of magnitude. They are based on statistics and extrapolation of experiments with animals. Reasonable consensus exists among workers in this field. However, in item 2, controversy is rather widespread, as our knowledge is still imperfect.

3.2.1 Electrical Hazards (60 Hz and Pulses)

Figure 3.5 extracts from the literature, in particular Refs. [1] and [3], a composite picture of the effect of electric currents on human beings. On the left the critical current (60 Hz for about 1 s) is plotted and on the right the critical energy of short (microsecond) pulses. The wavy line, actually a broad gray band, delineates the areas of potential death or mere discomfort. Each of the effects marked is actually a broad band, the effect being dependent on the weight of the person, the point of current entry, the size of the electrodes, the wetness of the skin, time, and so on. The worst situation exists for internal shock hazards: Less than 10 µA or 5 mW is stipulated for currents passing inside a patient's body, in particular in the chest area, because of the extremely dangerous ventricular fibrillation (quiver) that can occur (which we shall discuss shortly). Later we shall discuss the needs and methods for proper grounding and isolation of intensive care patients, as often quite unsuspected leakage currents can arise for internally wired patients.

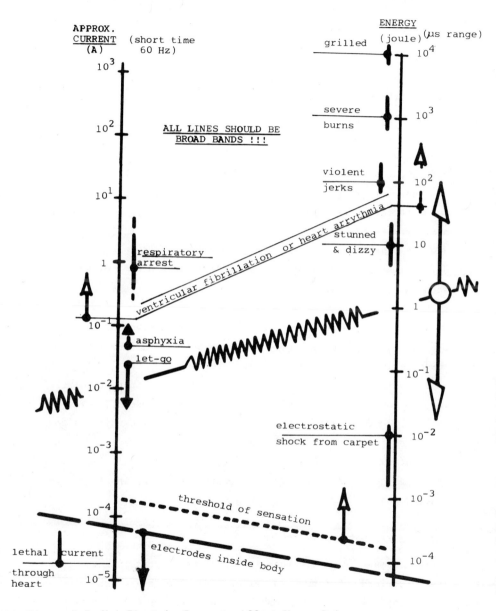

Figure 3.5 How Electric Currents Affect Human Beings

Turning our attention to external currents (no electrodes under the skin), we encounter quite a bit higher threshold currents: Between 10 and 100 mA is the borderline between "let-go" electrical current (the hand cannot let go of a wire) and asphyxia. Usually, respiratory arrest sets in at a somehow higher current level. Luckily, artificial respiration can remedy respiratory arrest and does not require special equipment not readily available.

As the resistance of the human body, limb to limb, is about 500 ohm, a 50 mA current is caused by 25 V. Hence, for a safe let-go current, the 60 Hz voltage should be well under 25 V.

Ventricular fibrillation causable by external current in excess of 0.1 A for a time longer than the heart period is a very serious problem. It requires the use, within 5 min, of a defibrillator to prevent death. For 5 ms, several kilovolts is needed. (If no defibrillator is available, heart massage should be applied in accordance with accepted techniques of cardiopulmonary resuscitation.)

The worst condition exists if a short impulse coincides with the most vulnerable peak of the T wave of the heart period. For the duration of less than three-quarters of the heart period, the susceptibility is about a decade lower than for longer exposure. In this context, we must warn again against the use of average values. A human life is at stake; thus the worst conditions count. Hence Underwriter Laboratories stipulates that 60 Hz leakage current must not exceed 5 mA. ANSI, however, has lowered this limit to 0.5 mA. High doses of current cause severe burns, and in the extreme case, death.

3.2.2 More Subtle Effects

There are many initially much less obvious effects than the ones just described. Yet such so-called side effects--accumulating over sufficiently long periods--may be highly consequential in terms of physical and mental health.

Human beings are much more complex systems than the most elaborate control systems they can build. Human beings have much larger computers (working both in the serial and parallel mode, with much larger memories), and they operate with great redundancy. Above all, human beings have consciousness, which human-made machines cannot duplicate. In the human body, electromagnetic forces of extreme subtlety govern highly interactive processes such that we often encounter only correlates instead of direct effects. But in spite of the exciting results of specific ingenious workers in this complex field, we still have no clear holistic concepts of human sensitivity to electromagnetic fields. An attempt to understand clearly life in conjunction with electromagnetic forces has so far been an elusive affair. It is as if we had herded many fishes into a pond; but when we try to catch them with our bare hands, they slip through our fingers.

Let us consider one example. What exactly do we know about brain waves? Well, we can delineate the ranges of alpha, beta, delta, and theta waves in terms of frequency and amplitude and correlate them with certain mental states. But how do we explain, for instance, that many acupuncture points can be detected by a marked difference in skin resistance, and that for people in the alpha (relaxed state) the resistance of these points changes even more? Now this is not idle talk but an important indication of delicate interactions that we can somehow describe but do not understand.

There is a pronounced--often extremely annoying--sensitivy of human beings to flicker. This flicker has a maximum effect in the very frequency range characterizing altered states of consciousness. In the range 0.5 to 13 Hz, a 1% change of amplitude of the voltage feeding an incandescent lamp is enough to be irritating. This disturbing flicker may be in the form of light, sound, or electric or magnetic fields impinging on a human being.

Now, alpha-wave feedback is useful for people who want but cannot learn to control their autonomous nervous systems in order to relax. It works because the person doing it feeds back his or her own alpha rhythm. But the situation is quite different if an external source with the frequency near the person's specific alpha frequency is forcing full or partial entrainment. The author used a small variable-speed motor to drive a disk having a slot by which light from an incandescent lamp was modulated. Sitting comfortably in his reclining chair, he changed the speed of the disks such that he came very close to his own alpha frequency. Really? Well, he knew exactly when he was close to his own alpha because the books on his bookshelf started to dance--at least so it seemed. The feeling imparted was initially rather pleasant and somehow soporific, so that he continued it until he went to bed. But alas, he had hardly fallen asleep when he was rudely awakened by frightening nightmares never experienced before--nightmares that made him run in panic out into the street in his pajamas. In toying with his alpha waves, he had entrained (see Fig. 3.1) or partly entrained his alpha waves, causing violent reactions. In experiments made at MIT, where people were subjected, without their knowledge, to electric fields near alpha-wave frequency, they were made highly uncomfortable. Modulation of energy in the range 1 to 13 Hz should be avoided because of its soporific and/or nauseating effect on human beings.

This little discussion on alpha waves has been given only as an indication of the unexpectedness of side effects. In most cases, resonances of subsystems or molecules are involved. The colloidal state may change. Sometimes simple heat effects are an explanation. The present state of knowledge in this fascinating and challenging field consists of some solid little islands in a vast quagmire. It is peculiar that the Russian and East European workers in this field seem to be much more concerned with functional and behavioral consequences of electromagnetic fields, whereas the U.S. and most European workers concentrate more on physiological and biochemical reactions.

Reference [4] is a survey of some of the unclassified Russian work. Much of their work is related to parapsychology. It is peculiar that the Russians appear to be much more open-minded than we are when it comes to working in such scientifically embryonic fields. This seems to be peculiarly contradictory to their otherwise rather rigid thinking.

References [5] through [8] are good surveys of exacting Western approaches to the biological effects of very low and very high frequencies. The emphasis on such frequencies got its impetus from the project Sanguine, now called ELF (communication with submerged submarines). On the other end of the scale, the radiation of microwave ovens, causing eye cataracts and, according to the Russians, even genetic changes, is still a rather controversial situation, which has fostered much more stringent regulations on microwave ovens. References [9] through [11] provide surveys over the entire frequency spectrum. Reference [12], appearing bimonthly, is an excellent magazine to read to keep up with the latest developments in biomedical engineering. We have included, from *Technology Review*, two excellent articles that look at the problem from a broad perspective [13,14].

REFERENCES

1. EMP Engineering and Design Principles, Bell Telephone Laboratories, Electrical Protection Department, Loop Transmission Division, Technical Publication Department, Bell Labs, Whippany, N.J., 1975.

2. Integrated Circuit Electromagnetic Susceptibility Investigation, Phase III, June 1977. IC Susceptibility Handbook, Draft 2, McDonnell Douglas Aircraft Astronautics Company-East (to be obtained from Cdr. U.S. Naval Surface Weapons Center, Dahlgren Lab, Attention: DF-56, Dahlgran, Va. 22448).

3. Lethal Electrical Currents, C. F. Dalziel and W. R. Lee, IEEE Spectrum, February, 1969, pp. 44-50.

4. Electromagnetic Fields and Life, A. S. Presman, Plenum Press, New York, 1977.

5. Psychophysiological Effects of Extremely Low Frequency Electromagnetic Fields: A Review, M. A. Persinger, H. W. Ludwig, and K.-P. Ossenkopp, Perceptual and Motor Skills, Vol. 36, 1973, pp. 1131-1159.

6. Uncertainties in the Evaluation of the Effects of Microwave and High Frequency Radiation, S. F. Cleary, Health Physics, Pergamon Press, Elmsford, N.Y., 1973.

7. Health Hazards from Exposure to Microwaves, Evaluation Group, Health Physics, Pergamon Press, 1975.

8. Special Issue on Biological Effects of Microwaves, IEEE Transactions on MTT, February, 1971, pp. 128-253.

9. Radio Frequency Fields: A New Biological Factor, J. B. D. Blanco, C. Romero-Sierra, and J. A. Tanner, Record of 1973 International EMC Symposium, pp. 54-59.

10. Overview of the Biological Effects of Electromagnetic Radiation, P. E. Tyler, IEEE Transactions on AES, Vol. 9, March, 1973, pp. 225-228.

11. Human Exposure to Non-ionizing Radiant Energy--Potential Hazards and Safety Standards, S. M. Michaelson, Proceedings of the IEEE, New York, April, 1972, pp. 389-421.

12. IEEE Transactions on EMB, bimonthly.

13. Signals and Noise in Sensory Systems, W. M. Siebert, Technology Review, May, 1973, pp. 23-29.

14. Electromagnetic Forces and Life Processes, R. O. Becker, Technology Review, December, 1972.

4 System Analysis
An Indispensable "Must"

In Chapters 1 to 3, we surveyed individual S's, T's, and R's. To prepare fully for controlling EMI sensibly, however, we must, in addition, analyze (for EMI) the system as a whole. This is not an easy task because the S's, T's, and R's are highly interactive and, in fact, often hidden. Moreover, the civilian systems we are here concerned with must compete in a cost-dominated market. In such a tough market, potential EMI costs, both short and long range, are not always clearly recognized, or if they are, the high price and limited effectiveness of conventional EMC make it appear a necessary evil to be dispensed with by a minimum of effort. But particularly in view of the proliferation of programmable logic controllers and microprocessors, the need for appropriate EMC is growing. The costs for system-commensurate EMC planning and implementation are much less than the costs, tangible and intangible, of an unhardened system, causing downtime, destruction, accidents, and lawsuits. Manufacturers of unsafe systems will lose their reputations and their businesses will suffer.

4.1 UNDERSTANDING THE SYSTEMIC DIFFERENCES

Military and communications engineers have established excellent, highly systematized analysis and prediction plans for their systems [1]. But their systems are fundamentally different from our systems: "Theirs" are essentially open-loop systems; "ours" are closed-loop systems (feedback, feedforward).

We cannot expect to achieve cost-effective electromagnetic compossibility of control systems by methods developed for communications systems. In communications systems and locating systems (e.g., radar), we have well-defined transmitters, channels, and receivers. In contrast, the indefiniteness and multitude of nonintended sources, transfers, and receptors and their tight intermixing with each other and with intentional sensor and control links make the EMI analysis of control systems more difficult. We must become clearly aware of these differences such that

We see the futility of applying conventional (communications-oriented) EMC methods and standards to control systems.
We can develop a system-appropriate method of EMI analysis and, hence, that
We can provide a congruous control method [always keeping in mind that the right question (analysis) is already most of the solution (control)].

Table 4.1 and Figure 4.1 give us first overviews to get us started in untangling this seemingly unmanageable mess. A control system consists of many interactive machine or process functions that often generate EMI (internally), to which external EMI, generated by near-neighbor systems, has to be added. The momentary state of each function or operation is measured by an (often very susceptible) sensor and transmitted via sensor channels (I/O devices included) to the automatic brain (EDP equipment) of the system.

The EDP equipment, centralized and/or partly localized (in the form of microprocessors), arrives at decisions about what to do and sends control information via O/I devices, being part of the control

Table 4.1 Generic Differences in Systems

	CHARACTERISTIC	COMMUNICATIONS SYST.	CONTROL SYSTEM
SENDER (S)	dominant	intentional, other information senders	incidental, switching transients
	bandwidth	narrow	wide
	density	low (mostly)	high
	timing	continuous	random
TRANSFER (T)	distance	great	small
	resonance	no	often
	radiated	far field	near field
	conducted	well defined and matched	complex and badly mismatched, spread
	mixed	no	very untransparent
RECEPTOR (R)	dominant	well defined (localized), tuned	the whole controller (sensor links, control links, & data processing equipment) is R
	density	low (mostly)	high
	bandwidth	narrow	wide
	amplitude	low	high
	hazards	hardly	always danger

generic differences

channels, to corresponding receptors and triggering devices. Now, although sensor and control links are designed as communication systems, they are tightly immersed in the noisy machine environment. This noise includes ubiquitous 60 Hz fields and currents. In this hostile environment, weak sensor signals must be enhanced by pre-amplifiers which may pick up strong interference via ground loops,

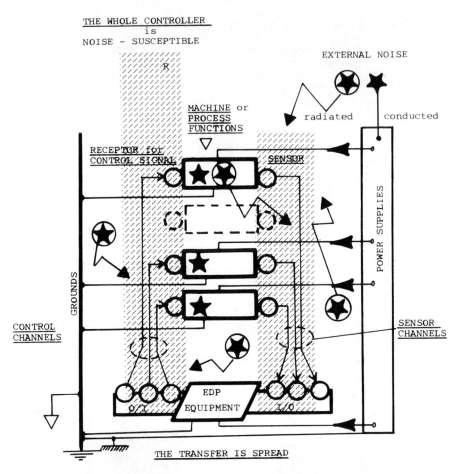

Figure 4.1 A Controller

mode conversion, power supplies, and so on. There cannot be an ideal ground since the system is spread over a large area. The situation is further aggravated if system ground, safety ground, and signal ground are not clearly distinguished. In ground systems, low-frequency and high-frequency effects are very different.

So what can we do to get a handle on this mess, where the whole controller becomes the receptor R immersed in a "sea" of broadband sources of high energy content and where the transfers are spread rather untransparently?

4.2 CRITICALNESS AND DECRITICALIZATION

Figure 4.2 portrays a first tentative scheme of the continual feedforward and feedback thinking needed for a progressive elimination of EMI. We shall see more of such heuristic thinking in Chapter 5 where--after repeated "massaging" of the presently still murky situation--we shall see a much clearer picture.

The term *criticalness* needs some clarification. Two aspects are involved:

1. We must determine the ultimate consequences of critical EMI effects.
2. After vicariously immunizing the whole system (through proper EMC packaging into silent, semisilent, and noisy spaces), we must identify those specific R's and T's that are by necessity immersed into the noisy space(s). These are the sensors and their amplifiers in the signal lines and the triggers in the control lines--all tightly immersed in the operating machines and processes constituting the noisy space. The noisy space is contrasted with the silent space containing sensitive computational equipment. The interface between these two is the semisilent space, which can tolerate some noise.

4.2.1 Analyzing for Consequences

In Figure 4.3 we consider just *one* receptor R and the 16 possible extreme consequences of its dysfunction, be it temporarily disabling or a permanent failure of a component or an injury. For simplicity, we stipulate further only N (not critical) and C (catastrophic) as extreme consequences. We are concerned--again highly simplified--only with (1) a localized consequence pertaining to a particular R *or* the human being exposed to this R (directly or indirectly), *or* (2) with a more or less complex chain reaction resulting in consequences for the whole system *or* with the disabled system affecting possibly whole groups of people.

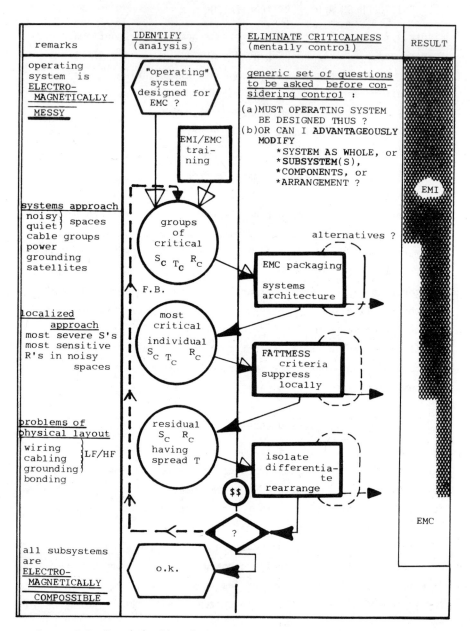

Figure 4.2 Decriticalization

Criticalness and Decriticalization 113

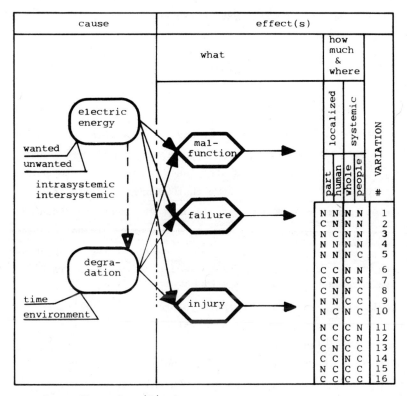

key : N = not critical
C = catastrophic

Figure 4.3 Analyzing for Consequences

In spite of the utmost simplification of our assumptions, we arrive at 16 variations of what can happen, both locally and systemically, as the consequence of *one* error only.

Variation 1 (NNNN). Nothing of consequence happens.

Variation 16 (CCCC). The one R is catastrophically affected and an operator nearby is badly injured; the whole system is a wreck and has plowed into a mass of people.

The 14 variations between these extreme consequences. For instance, in variation 2 (CNNN), R is destroyed, but there are no other consequences. In variation 5 (NNNC), a seemingly insignificant temporary malfunction of R does no damage to the system or its operator, but the system gets out of control, resulting in the death of a number of people. The system is still operative, though. Or what does variation 11 mean?

4.2.2 The Exposed R's

For the moment, let us assume that we have already EMC-organized our system such that we have delegated our most sensitive R's and T's to quiet spaces and our moderately sensitive elements to semi-quiet spaces. Let us further assume that we have isolated these two sets of spaces by more or less standard EMC techniques such as shielding, filtering, and grounding, such that they are not exposed to the disturbing sources which we cannot reduce in the noisy space. Yet there remain certain sensitive noise and hazard receptors R which are by necessity located in the unprotected noisy space of the installation. They must be there in order to measure or control the machines or processes constituting our system. Figure 4.4 illustrates this situation of necessarily exposed R's. They are exposed to coupling from intentional and unintentional sources of the noisy space (switching transients of actuators, RF sources, etc.). They are inundated by ubiquitous 60 Hz currents, and their harmonics, by stray currents, ground loops, or common impedance coupling, common mode coupling, and so on. The widely distributed ground and power feed systems transfer interference into the often rather sensitive exposed R's. What can we do about this? Well, in Chapters 5 and 6 we shall go systematically through options of making these R's into non-R's and T's into non-T's. Here, we shall only point out the broad generic principles because they are an immediate corollary of the foregoing analysis:

Use noncontacting and preferably nonelectrical sensors, as indicated in Figure 4.4.

Use isolators or light pipes to provide only nonelectrical T's. Amplifiers must be isolated with respect to input, output, and power supply.

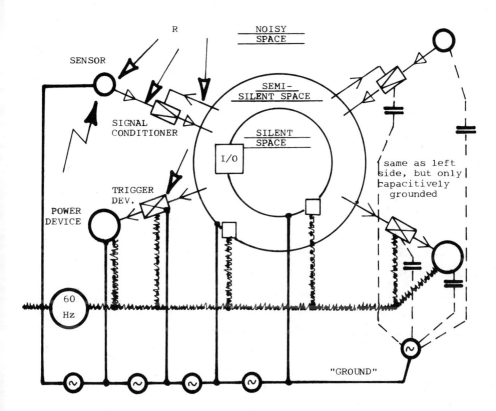

Figure 4.4 Exposed Receptors

We are now prepared to proceed with the many options of control, of preventing EMI and hazards. By grouping these various options under generic headings, we can develop a flexible strategy so as to be most cost-effective within the framework of the whole system. But before we enter control proper, let us delineate clearly what we intend not to do. We do not intend to teach system design per se, or standard reliability and quality assurance methods, or details of mandatory safety codes. Systems designers must know these things or they will soon be out of business.

REFERENCE
1. A Rational Basis for Determining the EMC Capability of a System, R. B. Schulz, IEEE Transactions on EMC, May, 1974, pp. 109-114.

5 Systemic Control
The Key to (Cost)-Effectiveness

Chapters 1 through 4 identified the essentials of analyzing interference in civilian systems: What can go wrong and why? Knowing the "why" is already a significant, though not a sufficient step toward control which we examine now. Admittedly, this is a complex undertaking. But we shall attempt not to fall into the common trap of treating sets of interrelated problems as unrelated and of oversimplifying difficult problems for the sake of tractability. Thus systemic control is the subject of the largest and cardinal chapter of the book. The reader should refer to the table of contents to see the plan of attack.

5.1 OVERCOMING HANDICAPS

Engineers responsible for EMC implementation face some seemingly obstinate managerial obstacles that may stifle their efforts. But these difficulties can be overcome once their real causes are recognized. Let us discuss first these handicaps and their circumvention.

5.1.1 Standards Are No Substitute for Thinking

We cannot dismiss the military approach to EMC without having another one that works. A simple modification of the military approach will not do. Rather, we are facing sets of real, highly interrelated, nonroutine problems. The key difficulty is that there are multiple objectives. For meeting each objective, there are multiple-choice means and bounds. Here the bounds are in the form of constraints, obstacles, and/or uncertainties. These are conditions usually not found in standards, except perhaps in a vague, general language of limited utility. Yet we ought not to complain about the present lack of enough standards appropriate for civilian EMC systems. We shall manage without them.

A holistic/heuristic method seems to be the reasonable way out of our dilemma. Step by step, we shall develop a comprehensive and practical approach to EMI/hazard control such that we get rid of vexing obstacles which we so nonchalantly disregarded in Figure 4.2.

Figure 5.1 portrays the first overview of such an attack on our problem. The straight path, from the task of the system to its operation, constitutes the dominant planning of the system as a whole into which EMC must be co-planned. The earlier we include EMC planning, the less is the total cost of EMC. This will become very clear once we recognize how crucial it is to select the proper system configuration and components (see Figure 5.1).

Heuristic means a self-educating method of systemic, yet flexible searching for reasonable solutions by continual feedforward and feedback. The heuristic method provides an aid to problem solving and requires thinking that is both good and flexible. In Section 5.3 we give basic general guidelines that allow us to come to grips with EMC in any spread system--by systematic partitioning and isolating interfacing. But you have to do the thinking and deciding in terms of your particular system. Such adaptive thinking is anathema to cookbook engineers who want a well-defined program for the computer such that they only have to plug in numbers, and the more so the better. But one cannot convert the interactive set of

Overcoming Handicaps

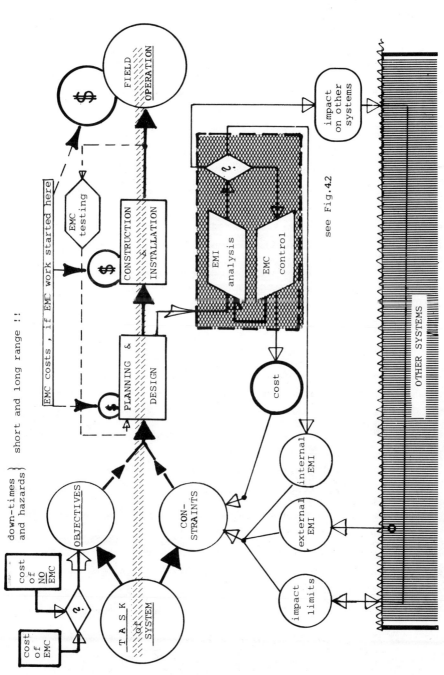

Figure 5.1 System Planning and EMC Planning

nonroutine problems into a routine affair. Or, looking at the situation from a different angle: It is very costly to employ rigid cookbook engineers for overall EMC planning. They are needed for support roles, however. The key EMC engineer should be the systems engineer. He or she has to study EMC, but not from "military" experts.

Well, that is enough philosophizing, which is often disliked but all the more indispensible for practicable results because theory makes things practical. But before we start with our quasi-systematic technical approach to EMC, let us briefly investigate another managerial handicap.

5.1.2 Secondary Role for EMC? Not If We Look at Costs!

The primary objective of the system is its proper and efficient function, with EMC often considered only a necessary evil of secondary importance. But, although admittedly only contributary, EMC, appropriately applied, can be the decisive contributor to the operational and financial success of the system, for two reasons:

1. The modest cost of proper EMC planning and implementation of the system can save considerable sums of money that might otherwise have to be spent sooner or later by embarrassing needs for retrofitting of systems or might be incurred by lawsuits occasioned by system dysfunction.
2. The need for EMC will necessitate a change of system architecture that may reduce system cost (e.g., fiber-optic "wiring").

As soon as the systems design management realizes the long-term cost of insufficient EMC work, it will take EMC very seriously. EMC is a contingency that may be integrated in systems effectiveness planning on a footing equal to that of reliability, quality assurance, environmental control, human engineering, and safety--all of them are "MUSTS." All these contingencies overlap with EMC--most of all safety.

In fact, the electrical safety code prescribes obligatory grounding and bonding such that we have to accept them as premises for EMC, even though they were not planned for. We have to take a careful look at such perplexing constraints.

Management needs two dollar figures to assess: (1) how much it has to spend for EMC in order to (2) prevent costs occasioned by EMI, which are to be estimated. This is a problem of optimization (ROI, return on investment) which is admittedly hard to quantify because:

1. We do not know how much it would cost if we ran the system without EMC provisions. Maybe the system would not operate at all.
2. EMC--as we shall develop it--will be so integrated into the system that the system configuration is starkly changed (different package due to fiber optics, etc.) such that the EMI-immune system is probably cheaper than the system not hardened at all.
3. As has been mentioned and will be investigated further, EMC and safety engineering are so closely interrelated via mandatory grounding and bonding that safety costs tacitly enter the picture.

To get a handle on this interactive cost situation, we start with something we are familiar with. For this purpose, we look at Figure 5.2--to begin with, only at the left side. The curves (a') represent the well-known minimax V curves for cost versus reliability of a system. Here we plot MTBC as the abscissa. Normally, that would mean mean time between failure (MTBF). But we change this to mean time between catastrophes, to make it commensurate with our treatment of EMC, to follow. As ordinates, we plot

1. The cost caused by failures (dotted) in the form of operating problems (i.e., loss of profit, etc., and repairs)
2. The costs we must afford to bring about reliability (dashed)
3. The sum of items 1 and 2 (solid lines) indicating a clear optimum

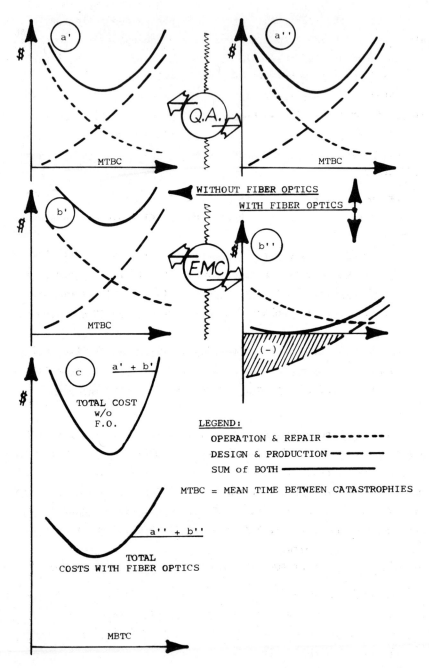

Figure 5.2 Costs versus MTBC (Mean Time Between Catastrophies)

Overcoming Handicaps

We see that total reliability costs are very high if we work with no quality assurance. They are also extremely high if we need extremely high reliability, as, for example, in space missions, where repair and maintenance are limited.

Now, the curves just discussed are based on accepted methods. In reliability engineering, we can quantify the probability of failure with a specified level of confidence, for individual components or parts, and can multiply these probabilities for all components to arrive at a probability of failure for the aggregate. Disregarding the burn-in time, we arrive at the reliability figure that the failure will not occur for a given time. Such conditional prediction is quite reasonable and acceptable for clearly delineated conditions, and a great deal of literature is available on this subject.

In Figure 5.2(b') we try to make a similar optimal consideration for EMC, including mandatory safety measures. We may consider reliability (a') as essentially "static" reliability only, and EMC (b') as a kind of "dynamic, operational" reliability. This analogy is not quite correct, as EMI functions are interdependent and multivariant such that the assumptions underlying a systematized prediction are a bit shaky. Linear programming will not help us to come to grips with nonlinear and interacting relations. A computer cannot help us to transform uncertain input data into certain output data. So what can be done to establish a reasonable creditability for a quantification of Figure 5.2 without committing economic suicide?

We shall come back (Sec. 5.2.1) to the basic question of why EMC, as defined in our context, cannot be handled primarily with mathematical prediction methods as they are meaningfully applied to reliability and communication-oriented EMC. But we can answer with a counterquestion: Isn't all business planning done under similar uncertainties? Certainly. Thus why reject carefully estimated costs for EMC when it is accepted for sales forecasting and, altogether, all business planning? This naturally presupposes that the forecasting and assessment team (going by the Delphi or some other accepted method) has applicable previous experience and possibly does some strategic experiments on critical subsystems. At any rate, we must admit that total cost estimation for non-EMC-immune systems is very much a matter of judgment, but it is easier to do so for hardened systems. The question now is: How do we know that we do not do too little or too much EMC? The answer: All effects of critical R's and spread T's must be made unconsequential. This can best be done by looking at the system as a whole. The upper curve (a' + b') of Figure 5.2(c) is the sum of the total cost of reliability and EMC and shows a very pronounced minimum. With the tentative assumptions and reservations stated before, we have combined (a' + b') the optimization of all costs attributable to technical failures, whether caused by unreliable components and/or EMC. Safety costs are temporarily assumed to be constant.

Now we go one step further, by anticipating that we already built the "ideal system" (partitioning and judicious choice of isolation), which we will arrive at in Section 5.3. Thus we now take a look at the right side of Figure 5.2. We find very much reduced costs, in fact so much so that negative EMC costs may result. This, with "ideal" EMC measures, means the cost of ideally-EMI-immunized systems is less than for systems having no EMC provisions at all, because in the co-planning process, the costs of the system proper were reduced.

We are now ready to plan our systemic control of EMI. We start with first things first, that is, with mandatory safety measures against abnormal conditions, in particular the worst kinds, lightning and ground faults. At the moment, we are not concerned with detailed specificities of safety measures; for example, we are not interested in what wire size to use for grounding conductors. Instead, our present objective is to delineate the generic effects of safety measures as they severely condition EMC planning for the system as a whole.

5.2 THE IMPACT OF "SAFETY FIRST!"

5.2.1 More About Reliability, Safety, and EMC in General

It is outside the scope of this book, but it is tacitly understood that a system must be designed for a high degree of reliability such that under normal operating conditions and for its lifetime (aging!), the system components will probably not fail. This is done by proper material selection, by taking into account the physics of failure, by providing sufficient creepage, by proper maintenance, and so on. Large systems would be highly unreliable if it were not for LSI (large-scale integration), whose use eliminates enormous numbers of solder or other joints which are very failure-prone.

For quite a variety of systems (e.g., hospitals or plants handling explosives or flammable materials) inherently hazardous conditions are "normal" conditions. The NEC, the UL, and others have standards to cope with such cases.

For any system, highly abnormal conditions may occur: too much energy or none at all. Excessive energy bursts must be diverted safely, or suddenly lost energy must be furnished by automatically switched-on standby power sources. For any abnormal conditions, the system must also be made safe: no fire or accidents. This is done by using circumspect design, by proper material selection (fireproofing), by providing protection devices, and other means. Naturally, all these precautionary measures must be introduced without prohibitive cost and excessive increase of size and weight.

Somewhere between normal and abnormal conditions lies the incidence of EMI. EMI is not a normal state of affairs, in that it is not always predictable and quantifiable, but it is normal in that it must be expected sometime and somewhere.

The preceding statement needs some elaboration. Statistical methods are applicable to normal-condition reliability calculations, as discussed before. Statistics are also useful in handling the occurrence of certain abnormal conditions if we can consider individual events (e.g., lightning) independently.

Example 5.1 The Likelihood of a Lightning Strike. According to Ref. [6], the effective strike area of a building is roughly the base area of the building enlarged by a strip twice its height, laid around its circumference. Statistics are available about the average incidence of lightning in different areas. Let us say that we have to consider the advisability of lightning prevention at a location where the statistics give the value of four lightning strikes per year and kilometer squared. For what structure is it advisable to apply lightning protection, and for what not? Let us take four cases:

1. A building 1 m × 1 m × 1 m high (a doghouse). The effective strike area is about 10 m^2. Hence we can expect one lightning strike in 25,000 years.
2. A building 12 m × 12 m × 5 m high (a garage). The effective strike area is about 1000 m^2, that is, one lightning strike in 250 years.
3. A building 25 m × 25 m × 5 m high (a house). The effective strike area is about 2000 m^2, that is, one lightning strike in 125 years.
4. A building 50 m × 50 m × 110 m high (a high-rise building). The effective strike area is about 250,000 m^2, or equivalently one lightning strike per year.

It is quite obvious that we do not have to provide lightning protection for a doghouse, even if it houses the most precious dog in the world and if the dog has diamonds worth millions of dollars in its collar. But it is also clear that we must protect the high-rise building against lightning effects. In this case, then, statistics are applicable.

But can we justifiably apply statistics to total EMC systems planning? The answer is no. Here we no longer have a simple zero-sum two-player game relating to a simple definable choice that permits determining the limit of diminishing returns. Rather, in EMC, we have an n-player game in which groups of players are in cohorts (chain events of EMI; see Chapter 4) such that statistics is of limited theoretical value, much less of practical value. Naturally, whatever we can gather in statistical information we shall use as we just did with lightning. But we cannot use statistics in closed form for the whole system as we can do for reliability and even for conventional communications-oriented EMC.

Our civilian EMC planning, then, must operate in just the opposite way in which conventional EMC is planned. Instead of considering individual sources, transfers, and receptors and adding individual effects, we must start with a generic "control plan" by partitioning and isolating whole groups of critical elements. In other words, our civilian EMC must be synthetic and not analytic, as military EMC can be.

Moreover, we have to start EMC control planning on the premise that safety codes have been met because they are the law and because they are interactive with EMC. The severest safety requirements arise for lightning and ground faults. Naturally, safety precepts for such high-energy problems were made by power engineers, who think essentially in low-frequency terms. Their sole concern is about taking the brunt out of these unavoidable energy bursts such that consequent further reduction makes safe possibly imperiled equipment or persons. Thus safety codes are not EMC codes; rather, they place constraints on EMC, the more so the more spread of the system, the faster the system is, or the more it is placed into a noisy environment.

5.2.2 The Relevant Essence of Safety Codes

In preparation for an investigation concerning mandatory but not always favorable effects of safety codes on EMC, we shall now condense very succinctly the pertinent points of the NEC as far as they influence EMC. And they do influence EMC--very much so!!

5.2.2.1 Lightning Protection

[We disregard here EMP (typical energy density 2 J/m^2; 10 ns rise time, 100 kV/m maximum; broadband with a frequency spectrum maximum at about 70 kHz) (see Refs. [1] through [6]), which is more severe than lightning in that it has a much steeper front and in that it is a distributed source instead of a point source, which lightning is. The ideal system developed later will contribute very much to hardening the system to EMP. And at any rate, EMP is a rather unlikely occurrence.]

Now, here is the gist of lightning protection codes.

1. In order to sink or divert the lightning energy into the soil, provide a lightning probe and ground it as prescribed in codes. Include grounding the water pipes, and so on.
2. In order to prevent side flashes, equalize potentials by bonding non-current-carrying metal masses. Include structural steel.
3. In order to prevent damage by lightning propagated along lines, use lightning arrestors at the input and lower-power voltage suppressors along the line.

In this context, we must mention briefly another critical case, that of static electricity. Its energy content is minimal, and no safety codes are yet written about it as far as I know. Static electricity is caused by friction, as for example, in the case of a person walking on a carpet. Static electricity is much worse in the winter when the air is dry. Up to 10 kV can be generated, which, in the equivalent capacity of a human being (on the order of 300 pF), represents $1/2\ CV^2 = 150 \times 10^{-12} \times 10^8 = 1.5 \times 10^{-2}$ J. Even a 1 kV charge can constitute enough energy to destroy integrated circuits or cause data errors. To prevent trouble that static discharge may cause in data processing equipment, it is mostly sufficient to use antistatic nylon carpets with conductive backing *and* to provide a relative humidity of 50 to 60%. To prevent static buildup in a manufacturing process such as in plastics or paper production (where static electricity attracts dust, damages products, may cause fire, etc.), commercially available electrostatic eliminating devices are applied. They are rows of little discharging needles (induction effect) or blowers of ionized air, if dust must be removed simultaneously. When carrying explosives, sliding into a car with plastic seatcovers or wearing a nylon shirt is a sure way to blow oneself up, particularly if the humidity is low.

5.2.2.2 Overvoltage and Overcurrent Protection in a Nutshell

1. In order to prevent 60 Hz leakage current from getting into the facility ground (double grounding of cable; see Sec. 1.3), ground 60 Hz cable shield at the facility entrance only. Use a separate ground rod.
2. In order to prevent excessive voltage imparted by switching transients and/or ground faults, ground the neutral (white wire) only *once*, at the entrance transformer secondary.
3. In order to interrupt excessive ground currents, use fast-acting protection devices such as circuit breakers.
4. In order to reduce shock hazards, provide safety wire (green) inside the conduit and attach to all cabinets, housings, and so on.
5. In order to prevent burnout and fires, use fuses and thermal cutoffs.
6. In order to prevent shock hazards, use leakage current interrupters, which must be responsive to small currents (5 mA); make sure that all the black wire (hot), neutral wire (white), and safety ground (green) are going through the current transformer of the interrupter (also for ground faults).

Some remarks pertaining to grounding the neutral are very much in order (Refs. [1] through [5]). In the United States, the code says "ground the neutral for the purpose of protection." Well, the main objectives are:

1. To prevent excessive voltages from oscillatory abnormal switching transients and the prevalent single-phase ground faults.
2. To have the three phases of the systems virtually independent of each other, for example, such that the phase switched off first does not present an extra-large load on the circuit

breaker. And as an important corollary: Feed noise-generating loads such as fluorescent lamps with phase A/neutral, and noise-sensitive loads with phase B/neutral (reduce transfer).

But this mandatory neutral-grounding causes a severe disadvantage, in that the ground fault current is very large. It would be much smaller and less destructive in an ungrounded system because there would be no return path (if we disregard capacitive distribution).

From the foregoing consideration, it becomes quite apparent that the obligatory grounding of the neutral may not be the best of the possible safety precepts. In fact [4], the United States and Canada, both of which require grounding neutrals, have bad records on electrocution, whereas, for example, Norway, which requires no grounding of the neutral, has one of the best safety records. There is some statistical proof, then, that alternative safety systems may be safer than the one presently accepted in the United States. Yet there are no indications that the code will be changed at the expense of another one which is apparently not quite so consequential a compromise. The electrical safety codes go into quite some detail. But still, without really understanding the underlying principles, a formalistic application of the code is not enough to assure safety.

5.2.3 Misconceptions and Antinomies Imparted by Safety Measures and Spread

5.2.3.1 Why We Have the Confusion

We come now to one of the most controversial and misunderstood areas of EMC: grounding and wiring in large systems. In them the simple approach applicable to small, slow systems, that is, "single-point grounding" and "no loops," is not realizable. Moreover, shielding, filtering, and mode conversion must be included in consideration of any wiring and grounding.

In this section we are not concerned with such specific aspects of shielding, filtering, and wiring/grounding that can be treated as separate problems. We discuss such disconnectable problems in

The Impact of "Safety First!"

Chapters 6 through 9. We shall see there that even such individual problems are often wrongly formulated by assuming unrealistic boundary conditions that are convenient but inadequate, hence very misleading.

Here, in contrast, we want to come to grips with sets of highly interactive problems caused by the all-pervasive wiring and grounding. Or more generally, our concern is to extract the essence of how the various conductors of a system affect systemic EMC. Such conductors include structural steel, water pipes, electrical power wiring, safety wiring, metallic cabinets, signal and control lines, and so on, all serving different purposes. Many contrary aspects are involved:

1. We build on Section 1.3 (please reread), where we considered relatively simple causes of counterpositive and anti-interference measures.
2. We include in our consideration also the mandatory--and unfortunately EMI-spreading--safety measures that we delineated in Section 5.2.2. And we have seen already that we had to make compromises there, too.
3. We include the consequential fact (which we must accept and cannot treat as if it did not exist as is done in many EMC texts) that we have a spread system that has large linear dimensions of wires and mandated cross-connections. The large frequency spectrum covered by EMI makes the spread highly significant, as the dimensions involved are often larger than the wavelengths of the operating and interfering frequencies.
4. In particular, we take a very suspicious look at the "control common," essentially a dc concept. It says: Form the grounding system into a tree; connect the control common only to the control cabinet housing the systems center (containing sensitive electronics). The signal return wire from other cabinets must only be hooked up to the system center (or system subcenter, if we branch out), not to all other peripheral cabinets. Well, if we have to introduce wire into a cabinet without

providing filtering, we invalidate the shielding of the cabinet. Thus, if necessary, we have to ground the control common, in terms of high frequency, through a filter capacitor. (Why must this be a feed-through capacitor?)

In Chapter 4, we talked generally about exposed R's and what could be done about them by isolation. But now we see that there are many more R's having T's (transfers) compounded by the mandatory safety wiring and ground we delineated in Section 5.2.2.

Admittedly, the problem of grounding and wiring is vexing if we work with a large system. To illustrate, we note that, for instance, the "IEEE Standard Dictionary of Electrical and Electronic Terms" contains 95 entries on grounding. Some people use different symbols for different grounding objectives. But this is essentially wishful thinking. Marking them as specific grounds does not make them so. Many papers have been written about wiring and grounding. Some of these papers are excellent as far as the delineated aspect of their treatment is concerned. The papers are written either by EMC people, who often disregard the crucial constraints imposed by non-EMC conditions (safety codes, etc.), or by power people, who stress safety and promote the concept of control common. (This is also called signal reference ground, signal return wire, etc.) As a power person might say: "If you use the control common, you already take care of most EMC problems. For special high-frequency problems, use shielding and filtering." Is that really so simple? No, the opposite is true, even if we avoid such obvious mistakes as digital and analog devices having common impedance via a power supply bus.

The reason for the grand confusion existing in grounding and wiring is the myopic way of taking things out of context. Here the context involves all FATTMESS criteria, as the reader will soon realize. The set of problems encountered is so central to EMC that we have to take a perspective look. To resolve the dilemma we are in, we survey (Table 5.1) the interactions caused by obligatory safety wiring and grounding *and* by the unavoidable spread in large and fast systems. Moreover, we discuss in some detail, by Examples 5.2, 5.3,

Table 5.1 Effects of Safety Measures and of Spread

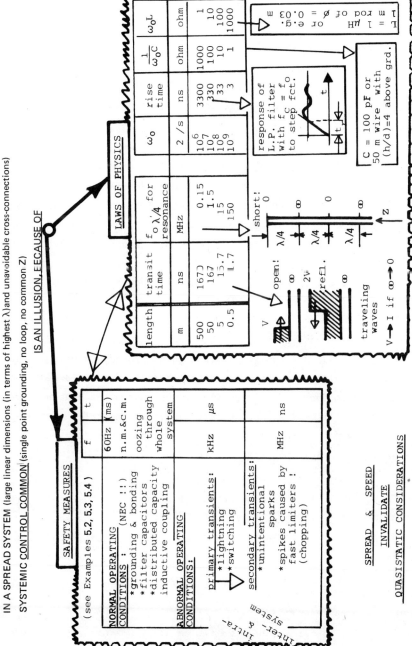

and 5.4, the most common misconceptions concerning the effects of safety measures; and we briefly delineate the effects of spread, that is, high-frequency effects, which are often not considered but are nevertheless very real.

5.2.3.2 Safety Measures Affect Normal Operating Circuits

Example 5.2 The 60 Hz Infusion. In Figure 5.3(a) we sketch two metallic cabinets, their power wiring, grounding, and bonding according to the control common concept. The green wire and the bonding wires are supposed to be normally free of 60 Hz. But they cannot be. Rather, the filter capacitors introduced against EMI, the unavoidable stray capacitances of wire runs, and the unavoidable multiple grounding effected by bonding causes 60 Hz stray currents to ooze into all conducting parts of the system. Loops are formed unavoidably. Normal mode and common mode are involved. In short, 60 Hz, even in normal operation, leaks through the whole system. Any two points of the grounded and bonded system can be represented as a 60 Hz source and an impedance. Hence for critical instrumentation, say in an intensive-care hospital room, all grounding and bonding should be made a localized virtual point ground, by using a cluster of receptables into which all electric devices are plugged, including lamps. Moreover, the bedframe, the instrument housings, *everything* metallic, is hooked up (preferably redundantly) to the safety (green) wires of the receptable cluster and are interconnected (intergrounding) such that we create an extended quasi-ground point (a ground plane).

In industrial systems, we cannot use this virtual zoned ground because of the large distances involved. Thus we have to rely on isolation and shielding, which also excludes magnetic coupling by stray fields.

5.2.3.3 Abnormal Conditions; Primary and Secondary Effects

We discussed oscillatory and critically damped switching transients in Chapter 1. Here we try to clear up some rather common

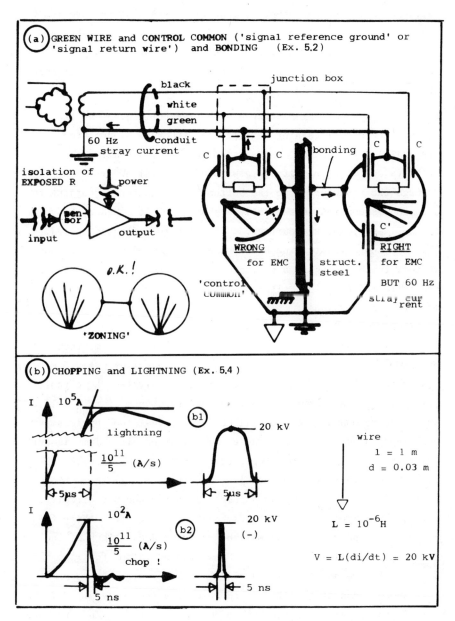

Figure 5.3 Sketches for Examples 5.2 and 5.4

fallacies about severe abnormal conditions. Again, the problem must be made tractable by simplification. But we must make the right simplifications and cannot take events out of context. To make sure that this is clearly understood, we discuss two examples (5.3 and 5.4) in some detail (Figs. 5.3(b), 5.4, and 5.5).

Example 5.3 Lightning Hitting a High-rise Building. The traveling-wave approach to lightning hitting the structure is depicted in Figure 5.4(a). It assumes a steep step function, say a 10^4 (10^5) A crest value, which results in a wave of 1.4 (14) MV traveling through the structure. At the bottom, we encounter, according to a common assumption, essentially a short circuit. Hence the voltage collapses, but the current doubles and is reflected upward. This is theoretically fine but does not jibe with reality because we took the building as a separate entity. Let us look at a more realistic model, which, although also very simplified, results in more realistic values.

In Figure 5.5(a1), we assume a flat, charged cloud 3 km in diameter, located about 2 km off the earth. Let us assume that a streamer has already ionized the air in a channel of d = 0.5 m toward the building. We can calculate the equivalent LC circuit [Fig. 5.5(b2)] and arrive at a resonance frequency of f_{equ} = 14.2 kHz. The resonance circuit will be severely dampened by losses contained in both the L and the C. The ionized air column is quite lossy, and the lower capacitor electrode (earth) is also heavily damped (eddy current in ground [see the term $(1 + j)$ in Section 7.1]). Thus roughly, we can assume that our resonance circuit is critically damped (see Table 1.2). Hence we plot from there $f_1(Q)$ for Q = 0.5 and mark the normalized rise and decay times as indicated.

From statistics we know that "reasonable" values for rise and decay times are about 5 μs and 35 μs, respectively. As shown, this corresponds to a critically damped resonance frequency of 12.7 kHz. Now, f_0 and f_{equ} are of the same order of magnitude. Thus we have reason to believe that our model is somehow realistic, although our approximations were very rough. At least we did not take the

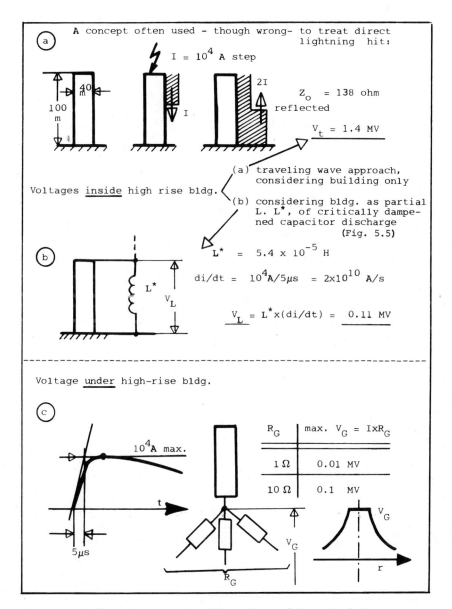

Figure 5.4 Pertaining to the First Part of Example 5.3

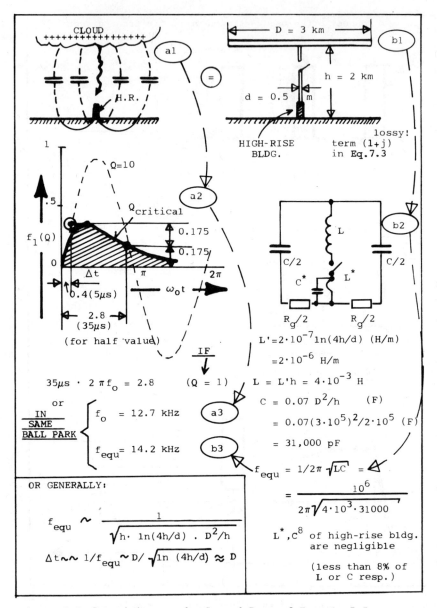

Figure 5.5 Pertaining to the Second Part of Example 5.3

high-rise out of context. Our main inference is: Lightning is a rather slow affair. Its slowness is not determined by the high-rise but by the whole discharge circuit (which we approximated crudely by an LC circuit). The L^x (and C^x) of the high-rise by itself is rather small in terms of the dominant L branch of the discharge path [Fig. 5.5(b2)]. In fact, we may disregard C^x. And going back to Figure 5.4(b), we can now calculate the inductive voltage drop along the high-rise to be 0.11 MV (1.1 MV for a 10^5 A peak) [see Fig. 5.4(b)]. This transient peak will last less than a few microseconds.

If the peak current is 10^5 A, we have 11 kV/m for a short time. Our electric system is in for a severe shock though the transit time for 100 m is only 0.33 µs.

Figure 5.4(c) considers what happens in the essentially resistive earth underneath the building. We should expect dangerous step voltages, which we can handle (see Chapter 9).

We return now to Figure 5.3(b) to discuss very fast secondary effects, namely current chopping, which often arise unintentionally, that is, by fast interrupting sparks and arcs (not only by contact opening but also by sparking in the neighborhood of lightning strikes or by the counterpositive effect of fast-acting voltage limiters). Such chopping may create pulses of nanosecond duration, normally on the order of several kilovolts.

Example 5.4 (Primary) Lightning and (Secondary) Chopping Juxtaposed. Since lightning is so much slower than chopping, the chopping of a current much smaller than a lightning current may create (in the same L) the same peak voltage as the lightning. In the example, we assume an L of 10^{-6} H to arrive at a spike of 20 kV. Intentional clamping is done by devices (e.g., Tranzorbs) mounted with much less L. But even an L of 10 nH would, in the example, create a 200 V peak nanosecond spike. Hence we have again a consequential counterpositive effect in that the high frequency contained in the chop causes radiation coupling between wires acting as antennas.

5.2.3.4 DC Considerations Invalidated by Spread and Speed

The right side of Table 5.1 illustrates over three decades of line lengths and frequencies, uncircumventable physical laws that we must take into serious account in fast-spread systems. Let's go from left to right.

1. Waves travel with a finite speed of 3×10^{10} cm/s for lines in air. That means that if we have a steep traveling wave, points close to each other may have greatly different potentials or they may have equal potential, depending on the time we make the observation. We indicated the reflection mechanism of traveling waves and assume that the reader is familiar with it, at least in principle, because we shall need this understanding when we treat test methods.
2. At high frequencies, a short in a transmission line is only a short if it is really a conductive plane. It is an L for a piece of wire across the line. But even if we have a perfect short, a quarter-wave away from the short we have an open circuit; or vice versa. An open line $\lambda/4$ long is a short circuit. For 150 MHz, $\lambda/4$ is 50 cm!
3. If we put a low-pass filter in the line, driven by a steep step function, the rise time behind the filter is made quite long; for example, for a cutoff frequency f_c = 150 kHz, the rise time is 3.3 µs.
4. In the first approximation, a long wire may present a considerable C and even a short wire may be a sufficiently large L to represent, at high frequencies, impedances very different from the open circuit for C and the short circuit for L, being the equivalent impedances for dc.

To summarize, we must accept the following facts (because of mandatory safety measures and of spread):

1. Sixty hertz oozes through the whole system because we cannot avoid loops and double grounding.

2. High frequency and transient interference has three different origins:
   A. It comes from the outside, by way of lightning or a switching transient, directly or propagated by power lines or water pipes.
   B. It is generated by the system operating under normal and abnormal conditions.
   C. It can be generated by EMC measures themselves.
3. Without isolation, the transfer of noise is rather unpredictable for the nonhardened system as a whole, as there are many redundant paths.
4. Grounding of the whole system (single point ground) is a myth, although we can establish localized quasigrounds, which hold only for the immediate neighborhood and which are different from each other.
5. "Control common" is a useful concept only for low-susceptibility, low-frequency peripheral devices operating within a confined space. Isolation or guarding is needed between localized grounds.
6. A short circuit for dc or ac is not a short for high frequency. The same applies to open circuits.

In the foregoing discussion, we have considered all FATTMESS criteria, with emphasis on interaction from the viewpoint of real and safe large systems. We are now nearly ready to plan systemic EMI control by partitioning and isolation.

5.3 EMC COMMENSURATE SYSTEMS DESIGN

Knowing engineering laws is not identical with engineering thinking.

5.3.1 Pertinent Trends in System Design

In Section 5.1 we investigated the managerial bounds; in Section 5.2 we delineated the technical bounds under which we have to do EMI control. The question is now: What can we do in view of so many interacting constraints? The answer is: (1) Use a holistic approach, and (2) if possible, circumvent, do not fight, constraints.

Whenever we are confronted with complex problems, we have to search for compatibility of the various *needs* and *means* (all combinations!) by continual feedforward and feedback thinking. See also the plan of the book, p. 000. In this mental process, we divide means into (+) remedies and (-) bounds. We have to check continually for means that have a multiple effect (desirable interaction) and for those that are counterpositive (undesirable interaction). A survey of these (±) means is provided in Figure 5.6 and in a bit more detail in Figure 5.7. We shall elaborate on this point soon.

With this in mind, we shall now, with a broad brush, marshal the significant opportunities feasible for EMC by present solid state developments still very much in flux. These fast-growing developments are basic--one could even say revolutionary--to system planning, more so even for EMC planning, as they permit EMI hardening of systems most effectively and economically. They provide the means to partition, to isolate, and to introduce redundancy, the very capabilities we need to extricate ourselves from the limitations imposed by spread and safety measures. The realms of information handling can now be completely separated among themselves and from the realms of energy handling, such that the information circuits are free of interference and hazards. Wanted and unwanted information can be made not to interact.

Here are these powerful means:

1. *Electrooptics*. Insulators replace wires for information handling. For this we have *sources*: LEDs (light-emitting diodes) or lasers; *modulators*: crystals of lithium niobate or tantalate; *receptors*: photodiodes, phototransistors; and *transfers*: (a) electrooptic isolators, eliminating the effects of double grounding and common mode coupling (such isolators can be made linear), and (b) fiber optics, eliminating all undesirable transfer and coupling.

EMC Commensurate Systems Design

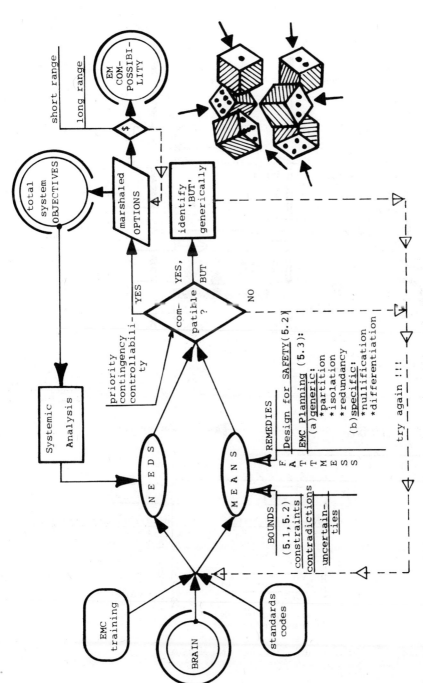

Figure 5.6 Overview of EMC Co-planning

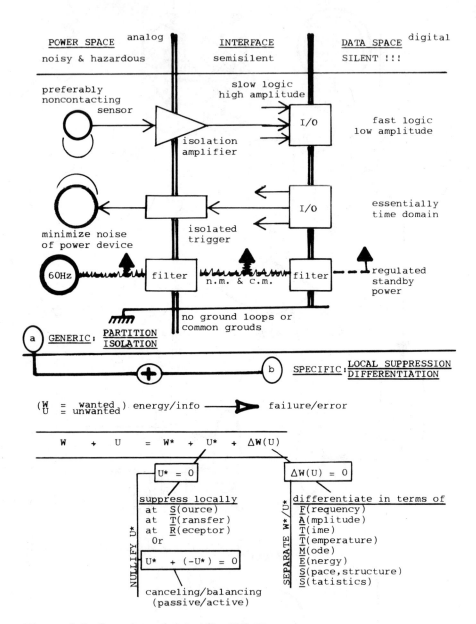

Figure 5.7 Generic and Specific EMC Measures

EMC Commensurate Systems Design   145

2. *Microprocessors* (μPs). Tremendous (programmable) wiring of logic at very low cost, in very small space, with very high reliability. Distinguish (a) one-bit machines (replace wired logic, decision oriented, from (b) multiple-bit machines (data logging). Networks (either bus or memory oriented) of (a), (b), or (a) and (b) approach computer capabilities. Hardware is replaced by software (see also Table 5.2).
3. *Charge-coupled devices (CCDs)*. Actually, these are shift registers for analog charge signals, in which the charges they store and transfer are introduced electrically or optically, free of switching transients. CCDs need no A/D and D/A conversion. Great reduction in digital hardware results. They also provide very fast scanning rates of linear and planar image sensors (much faster than vidicons). We have here sophisticated, non-contacting, self-scanning sensors that can also be used for signal correlating, signal reformulating, multiplexing, and digital filtering.
4. *Solar Cells*. Solar cells with a practical efficiency of about 10% are now available such that the power supply of preamplifiers in extremely hazardous situations can now be made wireless, thus harmless.

Thus, for the most part, we are now able to render the spread transfer harmless. From our FATTMESS attributes, the MESS criteria are most affected by the solid state technology development just delineated:

M (mode). Electrooptic permits the isolation of exposed R's, and electrical safety is vastly improved. The transfer of unwanted signals (common mode and converted normal mode) is nullified. Radiative coupling is zero. Resonances of lines, particularly critical for extreme terminations, cannot happen anymore (interfacial resonances).

E (energy). Software replaces hardware. With modulators of, for example, lithium niobate, fast data handling and multiplexing can be done in the realm of optics. Multidimensional Fourier

Table 5.2 Microprocessors

Objective	Automatic Control of Equipment		Large-Volume Data Processing
Input	ON/OFF sensors	Analog sensors	Data acquisition broadband (image) sensors
Complexity	One-bit programmable logic array (PLA)	Multiple bit μP, limited speed	Computer, high speed
Application	Decision- and command-oriented	Calculations and data logging	Complex calculations, pattern recognition, FFT, digital filtering
Language	Straight forward combinatorial logic (*familiar*)	Presently machine language (mostly *unfamiliar*)	Fortran and other languages
Expandability	Easy, just add; no preplanning necessary	Presently only if preplanned	Huge memories

transform is possible. For highly critical applications, the power of preamplifiers can be supplied via light beams and solar cells.

S (size). Size reduction (LSI) and decentralization (μP) permit quasi-single-point grounding in miniaturized subsystems. Single chips are available that operate directly with microprocessors, doing the multiplexing and A/D conversion. It will not be long before microprocessors and converters are one chip.

S (statistics). Systems planners can now afford to use full redundancy of critical signal circuits. They no longer have to restrict redundancy to limited measures such as parity checking. "Limited" is not meant as a nasty remark, but rather to contrast it with the affordability of EMC measures that were heretofore considered extravagancies and out of reach.

Admittedly, electrooptics and microprocessors are not an automatic cure-all for EMI, but they make things much more manageable and affordable (wiring greatly reduced or made nonwire) for information-handling circuits. In particular, electrooptical solutions have some limitations, which we discuss in Section 5.3.3.

5.3.2 Generic Remedies

5.3.2.1 Automated Safety Protection (Fig. 5.8)

We already discussed mandatory safety measures and how they affect EMC. We accept them. With respect to the critical attribute T (temperature), we have at our disposal thermal cutoffs to protect electrical and electronic products from overheating. And we tacitly assume that the system designer is familiar with the customary protection devices of power-handling systems. By fiber optics, we can completely (see the ideal system later) isolate all information-handling subsystems and their interconnections such that we do not have to concern ourselves with their safety from lightning and switching transients. Such nonconcern, however, is justifiable only if we include the power supply in the isolation.

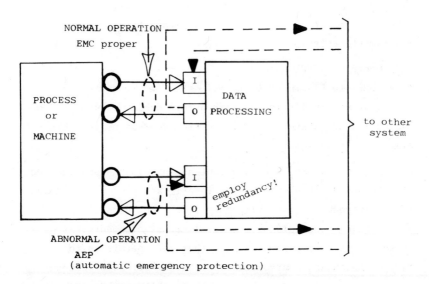

Figure 5.8 Automated Shutdown

Still, we have to indicate briefly two programmable protection schemes that are needed and that have become highly reliable and economical (and they are all commercially available).

1. For smaller systems, it is advisable to use automatic switching-on of isolated standby power for sensitive electronic equipment in case of approaching thunderstorms. This can be accomplished by charge sensors (little lightning rods). If the currents collected from the atmosphere exceed a certain value (depending on the sensors), the operation is automatically switched on standby power, thus creating isolation from abnormal conditions (a typical supplier: Lightning Eliminator Associates).

2. For complex systems, more elaborate protection schemes are needed. Microprocessors permit an orderly and reliable shutdown (and re-startup) to protect economically against catastrophies that can be caused by system failure, out-of-limit conditions, operator mistakes, EMI, external threats, and so on. With sufficient strategically placed sensors, microprocessors permit redundancy and sophistication such that only consequentially critical events are acted upon. Protection is then provided by means of hardware and software (a typical supplier: Rochester Instrument Systems).

5.3.2.2 Generic EMC Planning

We now go back to Figures 5.6 and 5.7. We find that the spread transfer that originally looked like a formidable problem can now be handled with relative ease. There are two reasons for this sudden clarity of relations:

1. We now know what we have to accept as unavoidable and what we have to avoid, thanks to our systematic analysis and reanalysis (after introducing the obligatory safety measures).
2. We now have at our disposal several new developments, as discussed, that are so far-reaching that they permit us to take the confusion and unpredictability out of spread transfer. In fact, interference is now rather confinable to the power devices and the power feed lines.

Here then, is what we can do systematically to provide cost-effective EMC:

1. *Partition of system into:*
    A. Quiet spaces containing low-amplitude, fast logic. Logic not faster than necessary!
    B. Semiquiet spaces, the interfaces containing slow, high-amplitude input/output devices. Make the speed not greater than necessary.
    C. Unavoidably noisy spaces, containing, in addition to power devices, the exposed receptors.

    The partitioning is facilitated considerably by microprocessors, which permit establishing distributed subsystems, which can be partitioned again.

2. *Isolation* (see also Chapter 6)
    A. By shielding and filtering of quiet and semiquiet spaces.
    B. Of actuators (control elements) by electrooptical isolators.
    C. Of exposed R's of great susceptibility by using
        (1) Noncontacting sensors (see Sec. 5.3.2.3) or electrooptical isolators.
        (2) Isolating amplifiers with isolation for input/output *and* power supply.
        (3) Fiber optics for critical exposed transfers. The radiative coupling into such transfers is absolutely zero. But, naturally, there are still the EMI problems of normal mode errors (or failures) affecting the electrooptical transducers themselves. We still have to resort to conventional EMC measures to immunize the electrooptical transducers. Nevertheless, we can be sure that lightning and 60 Hz ground faults, and even ground stray currents, find transfer $T = 0$ in fiber optics.

    Electrooptical isolators and fiber-optic devices are rapidly evolving hardware. Their manufacturers often supply detailed application notes on their products. Thus we can confine

ourselves to the principles and pitfalls involved (see also Chapter 6).
3. *Redundancy*. Redundancy (as already mentioned) of data handling subsystems is now economically feasible. Redundancy may be either duplication or triplication of whole critical subsystems or may be error-correcting code. References [10], [11], and [12] are excellent tutorial papers on this important subject. Thus error and fault tolerance can be programmed into the system by redundancy of the data handling subsystems and by software.

For the power handling subsystems, redundancy would be too expensive. There, redundancy must be replaced by subdividing of power aggregates such that emergency operation is still feasible when one of the power aggregates fails.

5.3.2.3 The Ideal System

In Figure 5.9 we summarize our findings thus far; we sketch the ideal and the nearly ideal EMC-commensurate basic system block. Only a few additional comments are necessary, namely: Three categories of noncontacting sensors are at our disposal:

1. *Digital noncontacting sensors*. These range from noncontacting position sensors or counters based on interrupting light beams or ultrasonic beams to noncontacting card or strip readers, and in some cases, exploitation of the Wiegand effect is advisable.
2. *Analog noncontacting sensors*. Capacity, eddy currents, or Hall effect may be the basis of displacement measurements. Photonic Sensors (MTI, Latham, New York 12110) use ingenious patterns of fiber optics to make fast, noncontacting displacement measurements of extremely high resolution (10^{-6} cm at best). And many parameters can be converted into displacement. Even the temperature of moving products can be measured without contact by the principle of convective null-heat balance [13].
3. *Image sensors*. The most sophisticated sensors are based on CCDs. They are line scanners presently up to 2^{11} photodiodes or area scanners (100 × 100 matrix) with scan rates in the

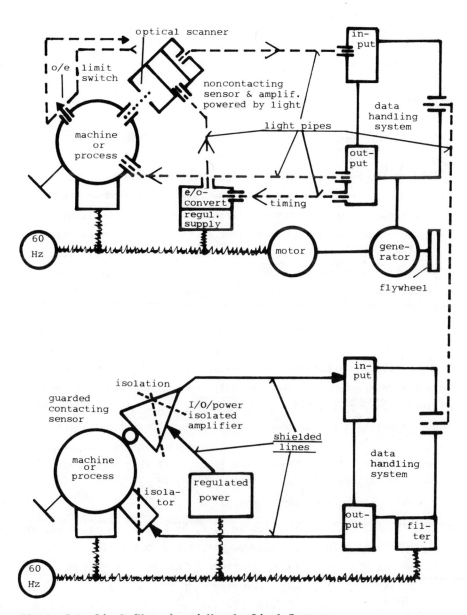

Figure 5.9 Ideal (Upper) and Nearly-Ideal Systems

megahertz range. They can be used for imaging, pattern recognition, and so on.

If noncontacting sensors are not feasible, we can always resort to electrooptical isolators or common-mode guarded (all three ports isolated) amplifiers. Manufacturers supply detailed application notes.

5.3.3 A Perspective View of Conventional Specific EMC Measures

Besides the broad and new strategies made feasible by the developments just discussed, there are many long-established specific means that should be part of the EMC system planning. We certainly need some of them, in particular as long as the development and mass production of the new devices is still very much in flux and hence presently in some cases expensive, and/or as long as they impose performance constraints (e.g., poor linearity, limited dynamic range, poor life). Let us look at these conventional EMC measures from as many angles as possible such that our problems appear--from one of these angles--in their greatest simplicity, so that we can EMI-harden our system with a minimum of cost. Naturally, we must assess the various options carefully for contingencies. Figure 5.6 gives the general plan of thinking we may follow, whereas Figure 5.7(b) marshals the broader aspects of such heuristic thinking.

Wanted (W) and unwanted (U) energy and/or information propagates and is modified through the system such that we encounter U^* and ΔW (U) as failures or errors. For solutions, we have three general approaches at our disposal:

1. $U^* = 0$. Local suppression at S, T, or R (source, transfer, or receptor). This is discussed briefly in Chapter 6.
2. $U^* - (-U^*) = 0$. Here we consider balancing or canceling, both actively and passively (see also Chapter 6).
3. $\Delta W(U) = 0$. To avoid falsification of signals or breakdown of equipment, we differentiate U and W in terms of the critical attributes FATTMESS. Although we shall come back to some of

the pertinent topics in later chapters, we now look at specific EMC means in terms of various options at our disposal as they affect the planning (and cost) of the entire system.

Figure 5.10 surveys some of the most important and practical FATTMESS options available to make $\Delta W(U) = 0$. They are given as a stimulus and as an indication of the flexible and yet rather systematic thinking that forms the basis of any good engineering. To start, we make the broad distinction of *direct* and *indirect differentiation*. By *direct* differentiation, we mean staying within the realm of a particular FATTMESS parameter. For instance,

For F. If W is a low-frequency signal and U is a high-frequency noise, we use a low-pass filter.

For A. If the W is a low-amplitude signal and the U is the short, high-amplitude spike, we use a limiter (MOV or clipper) to get rid of most of the disturbance.

For T_i. If the W is disturbed for a short interval, U, we eliminate U by blanking the signal during this interval. And so on for ...TMESS as given in Figure 5.10, ending with correlation techniques as indicated.

But quite often the distinction within one realm is rather expensive, blurred, or difficult in view of certain crucial system requirements. Then we resort quite effectively to multiple-criteria or *indirect* differentiation (lower part of Fig. 5.10).

The columns (a to e) of boxes compile the most important indirect EMI remedies. The boxes are connected to different FATTMESS parameters by different lines to show their relations. Note that, unavoidably, there is some overlap and quite some incompleteness in our presentation. Such imperfections may annoy rigorous perfectionists. But this course is not written for those engineers who like to categorize problems by one single principle of classification. That method is easier, but not adequate, in that it restricts possible facets and thus may discard the solution that is most appropriate under given concomitant circumstances. On purpose, rather, do we look at our interactive problem sets (not a singular

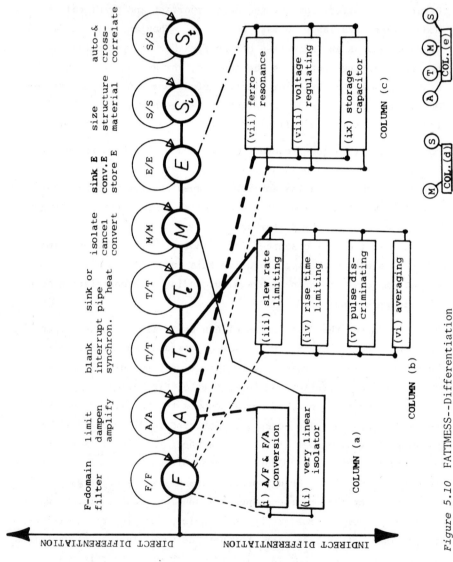

Figure 5.10 FATTMESS--Differentiation

problem) from many perspectives as indicated in the feedforward/
feedback diagram of Figure 5.6. There, we put into operation the
the best computer there is--the brain--operating in the intuitive
mode. From the alternatives thus established, we always can determine the most cost-effective solution by technical computers.
But to repeat, to get a list of alternatives, we must think adaptively, not routinely. The shaded insert in Figure 5.6 symbolizes
"look at all facets!" And we may call what we will be doing now:
Facetting!

We now go into some important details of indirect differentiation (lower part of Fig. 5.10), for which Table 5.3 provides
sketches facilitating brief explanations.

Column (a): V/F and F/V Conversion (i, ii; see also vi, viii)

The conversion of $\underline{V}$oltage or more generally $\underline{A}$mplitude, to
$\underline{F}$requency corresponds very much to the great advantage of FM over
AM [14,15]. Actually, we have here a digitalization in that an
analog signal is converted into output pulses of constant width
and amplitude, with a pulse repetition rate linearly proportional
to the input amplitude. No synchronization is required or permitted. The output is DTL and TTL compatible. Typical applications are A/D converters of high resolution, long-term precision
integrators, and two-wire high-noise-immunity digital transmission.

The converse, F/V converters, by themselves are very useful in
speed controllers, frequency monitoring, VCO stabilization, and so
on, since they are so highly linear.

Such converters are very economical, and their combination permits two-wire analog-in/analog-out transmission of high noise
immunity, both for normal mode and common mode. As most manufacturers of such converters (e.g., Analog Devices, Burr-Brown, Datel)
provide excellent application notes, we do not have to go into details. But we must stress the following important points: (1)
Advantage: Very large dynamic range; (2) Advantage: Excellent
linearity combined with high noise immunity makes V/F converters
exceed those of linear electrooptical isolators to a large degree;

Table 5.3 Sketches Needed to Explain Figure 5.10

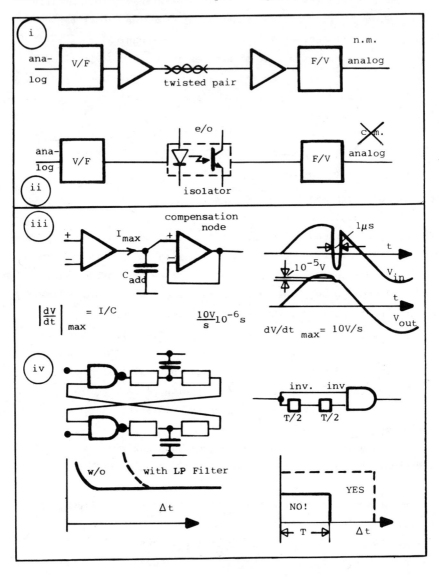

Table 5.3 (continued)

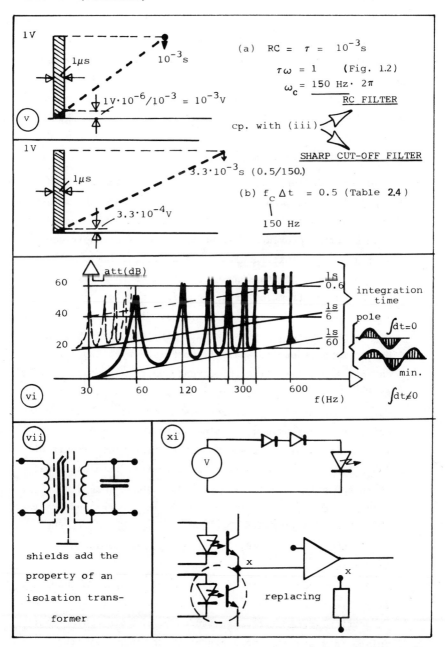

and (3) Disadvantage: Asynchronous, hence only useful in synchronous systems if V is reconstituted by F/V converter.

Column (b): Multiple-Domain Timing Effects (iii to vi)

Essentially all digital computing is done in the time domain. Thus impulsive interference can often be effectively eliminated from the time domain directly. Yet often the indirect approach, that is, including other FATTMESS parameters, is preferable. Let us look at four of such possibilities, based on different principles to be chosen as dictated by concomitant conditions.

(iii) *Slow rate limiter*. This is actually a nonlinear low-pass filter limiting dV/dt. Such devices are commercially available (Non-Linear Filters), but they can be built quite easily from op-amps, provided that we have access to their compensation point. This results in an excellent low-pass filter for low-power applications, permitting us to erase sharp random pulses from low-frequency signals. In contrast to a linear low-pass filter, the nonlinear filter does not introduce a phase lag. Thus slew rate limiters are often much better than linear low-pass filters (see v). Typical applications are eliminating the strong tracer pulse of pacemakers from ECG tracings or the wiper noise from potentiometers.

(iv) *Pulse discrimination*. If one wishes to pass a longer pulse while invalidating a disturbing short pulse, one can stay in the time domain, as shown in the right-hand circuit of Table 5.3 (iv): One inserts two inverters into one input of an AND gate such that only pulses longer than $2 \times T/2$ are transmitted. Or doing it indirectly in the frequency domain, one can make a flip-flop respond only to long pulses by putting low-pass RC filters in the outputs of cross-coupled NAND gates.

(v) *Linear low-pass filters as rise time limiters*. Although slew rate filters (iii) are much better than linear filters, in particular for large amplitudes of the disturbing short pulse, they are also more expensive and power-limited.

Assume a 1 μs 1 V (or 10 V) rectangular pulse. Use (a) an RC filter ($\tau = 10^{-3}$) or (b) a sharp cutoff filter with f_c = 150 Hz ($\omega_c = 10^3$). In case (a), the pulse is reduced to 10^{-3} V (10^{-2} V); in case (b) to 3.3×10^{-4} V (3.3×10^{-3} V). Compare this to (iii), where we did not have to consider the amplitude of the pulse; we had only to make sure that the maximum slope of the highest passing frequency is below 10 V/s.

(vi) *Averaging.* Averaging or integrating can be an excellent method for normal mode filtering, in fact far superior to frequency-domain filtering, if and only if the U (unwanted) is a continuous, nearly stable frequency (and harmonic!). Here the U may have a much higher amplitude than the W (wanted signal). Or more specifically, averaging is best for power line (60 Hz) related noise. For an integrating time T_a, all frequencies n/T_a (n = 1, 2, 3, ...) have infinite attenuation. Such poles are rather sharp, while the minima increase with 20 dB/decade for increasing frequency (explain).

Column (c): Notching in power lines (vii to ix)

Here we are concerned with sudden, short drops in the supply voltage due to suddenly increased load. Energy storage elements must be applied to supply the sudden energy demand. (This is the opposite of limiting, which we apply when we suddenly have too much energy.) Notching is best reduced at the source by using a stiff source or applying compensation and/or filtering at power converters (see Chapter 6). To remove notches from power lines, the following economical devices are available:

(vii) *Ferroresonant transformers.* These transformers utilize resonant LC circuits as an energy storage element for ac (f_0, ±5%). One can envision this circuit as a double-degenerated parametric amplifier where pump, idler, and signal frequencies are identical. Such devices are very rugged. They typically regulate the voltage within ±1% for an input voltage change of -20% to +10% and within ±2% for a load change from zero to full. Good ferroresonant transformers have enough energy at their disposal to supply a half-cycle of 60 Hz with the missing cycle caused by a lightning arrester or ground fault portection devices. Ferroresonant transformers are also good normal mode and common mode (with a Faraday shield) filters, the best having about 50 to 60 dB attenuation, at least above 1 kHz. Depending on their configuration, their performance varies widely. Manufacturers' literature should be carefully checked for excessive reactive current and for failure to recover from excessive notches.

(viii) *Voltage regulators.* Voltage regulators, be they of the series or shunt type or of the switching type, are described in the manufacturers' data sheets. In contrast to the high efficiency of the ferroresonant transformer (85%), switching regulators have about 60% and series regulators about 40% efficiency, a price one has to pay for better regulation and for much lower magnetic leakage fields, which can be quite disturbing for ferroresonant transformers.

(ix) *Storage capacitors.* Storage capacitors of sufficient capacitance are the most convenient--and often overlooked--elements to prevent notching in dc lines. Only in severe cases are active notch filters indicated.

Column (e): Temperature and Amplitude Effects in Electrooptical Isolators

(xi) *Temperature compensation of LEDs and diode-electrooptical system LEDs*, driven by a constant voltage source, emit less light at higher temperature. This can be compensated by putting a few small-signal diodes in series, as they create less voltage drop at higher temperature. Similarly, replacing the load resistance of a phototransistor, as shown, greatly reduces the temperature sensitivity of the coupler.

EMC Commensurate Systems Design   161

(xii) *Linearization of electrooptical isolators* (no sketch).
Electrooptical isolators may be operated in the pulsed
(digital) or analog mode. For the analog mode, the LED/
phototransistor types are too nonlinear. But they can be
made reasonably linear by using pairs compensating each
other. In contrast, the LED/diode type is inherently quite
linear, but the current transfer ratio is poor: mA's result
in μA's. At any rate, manufacturers' specification sheets
should be consulted, in particular with respect to the
dynamic range of linear behavior.

We conclude our brief discourse on indirect EMC measures with an example.

Example 5.5 Ratiometric Measurements. Suppose that we have a resistive sensor element in one arm of a resistance bridge. A large dynamic range of 10,000:1 is required, but the reference voltage V_R feeding the bridge is varying widely (U!). What alternatives do we have to solve this problem? We can

1. Use a voltage regulator for the bridge voltage. But this would require an extraordinary effort.
2. Use a digital voltmeter having ratiometric provisions. But normally this would only be sufficient for about ±10% variation of V_R and would require costly A/D converters. But we can (and will)
3. with reasonable cost and excellent performance, make ratiometric measurement with the aid of two V/F converters as sketched in Figure 5.11. The signal V_s is sent through V/F converter 1 to a counter whose time base is determined by the bridge voltage sent through V/F converter 2 and a digital divide-by-N circuit. Details of such a circuit are described in Datel's "Designer's Guide to V/F Converters."

Example 5.5 has been grouped under indirect EMC measures (because we changed an unstable A (V_R) into a stable A ratio via _fre_-quency, _t_ime, _m_ode, and _s_tructure). We could as well have put the

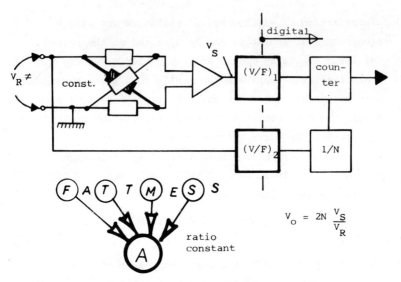

Figure 5.11 Ratiometric Measurements

example in Chapter 6 as a nullifying or compensating means. Remember "facetting."

5.4 COMMENTS ON SPECIFIC CLASSES OF CIVILIAN SYSTEMS

What we have discussed thus far referred to nearly "any" system. We turn now to codes and guidelines on safety/EMC-commensurate design of specific groups of systems, classified as follows:

1. Industrial systems
   A. Machine tool
   B. Process industry
2. Transportation systems
   A. Mass transportation
   B. Automotive industry
3. Commercial and professional system
   A. Buildings containing EDP equipment
   B. Health care systems

However, the information we are after is often rather fragmented and incomplete. Many standards are still in the tedious process of being formulated. To stay current, the reader should contact the various societies and associations (e.g., IEEE, SAE, EIA, ISA, SAMA, ASTM). Reference [17] lists the name of many a committee chairperson to be asked for information on pending revisions and new standards in the works. The same source also provides surveys on Military, FCC, and International Standards. But we have only occasional use for such standards, as we can scarcely design and test our civilian systems by Mil. Std. 461. Comments on pertinent standards are given in Chapter 10 and Section 12.2.

Again giving priority to safety, let us start with the broad codes and guidelines that are applicable to most of our systems:

1. The NEC (National Electrical Code) [18]. In this context we also mention the important UL (Underwriters' Laboratories, Inc.) standards, concerned primarily with specific products.
2. Recommended Practice for ([19] through [23]). Color books of the IEEE concerned with power systems and grounding.
3. ANSI Standard C39.5-1975: Safety Requirements for Electrical and Electronic Measuring and Controlling Instrumentation (SAMA and NEMA) [24].

For specific systems classes, however, we must include pertinent key papers to obtain the most comprehensive coverage presently available. Thus we have

1. *Industrial systems*. References [25] to [28], all of limited scope but complementary to the "any"-system approach set forth in Sections 5.2 and 5.3 which were, in any case, heavily slanted toward industrial systems. See also Figure 5.12 for a pictorial summary.
2. *Transportation systems*. Here we find more information on EMC on the systems level. In particular, the 1973 (and 1976) Records of the International IEEE EMC Symposia, [28] to [33], contain many articles on EMC in mass transportation systems. SAE Committee AE-4 is very proficient and has many standards under development. So it is best to contact its chairman,

164   5. Systemic Control

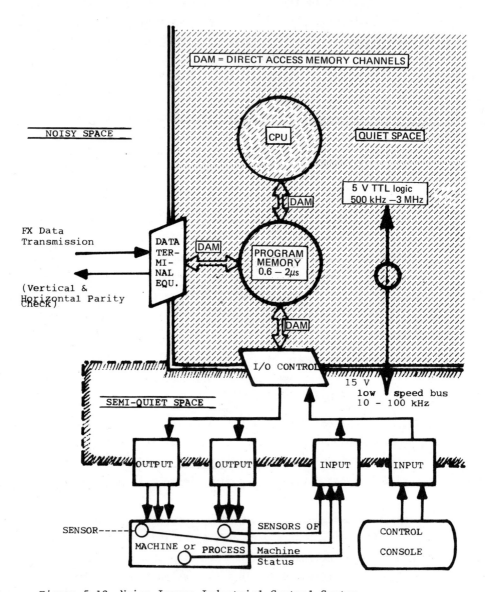

Figure 5.12 Noise-Immune Industrial Control System

presently Mr. Moe, in particular with respect to the work of
the SPECIAL TASK FORCE for Rapid Transit Systems. Concerning
the automotive industry, the reader should refer to Refs. [34]
to [36]. These papers are a few years old but still very pertinent, as they portray the basic problem areas which are still
dominant, although fiber optics and μPs are not discussed.

3. *Commercial and professional EDP systems.* Here we are concerned
with two subclasses:

   A. More generally, the electrical *problems facing the architect* of buildings containing EDP equipment. This critical
   problem, which is very expensive to correct retroactively,
   was discussed in part in Section 5.3. Our comments will be
   concluded and summarized in Chapter 9, where we also cite
   additional key bibliography.

   B. Particular aspects of the other subclass, *health care centers*, were met in Sections 3.2 and 5.2. In these potentially deadly environments, patients are often wired
   directly into their electrolytes. Life-saving attempts
   may become life-taking. Extreme precautions for isolation
   are necessary. For a more detailed introduction to the
   sensitive, fluid field of patient care, we cite three key
   papers: R. K. Jurgen's paper [37] is an excellent discussion of the Medical Device Safety Act and its ramifications.
   The Hewlett-Packard paper [38] gives a clear explanation of
   the causes, effects, and control of electrical shock,
   although the paper (1971) does not consider electrooptical
   isolation and current-limiting diodes. A Medical Device
   EMC Standard [39] is being prepared for the FDA, pertaining
   to EMI emission and susceptibility of medical devices.
   This is a first attempt to define the electromedical environment acceptable for different locations and equipment
   classes such that the necessary shielding and filtering
   can be determined.

5.5 IN SUMMARY

By considering the system as a whole and by introducing partition, isolation, and redundancy, we have tamed what originally seemed to be an untameable "can of worms." We must reconcile the rigidity imposed by standards and the flexibility needed for technical progress and cost-effectiveness. We can manage it if we *USE THE CODE AND USE THE HEAD*.

For the rest of the book (Chapters 6 to 12), our considerations will be simpler and essentially amendatory. That is: We have no intention of setting forth a complete, methodical treatise on conventional "very simple" EMC measures, which are supposedly well-understood engineering principles. So why should we discuss simple EMC remedies altogether? "Because"--says the sad engineer--"they often do not work." Our objective, then, is rather limited, but all the more important. Investigate why they do not work in spite of their simplicity and well-established theory. We must look at "simple" EMC measures with a very critical eye to make sure that we do not overlook any crucial FATTMESS criterion that might invalidate an oversimplified, unpredigested concept. These simple elements have to operate in a not-simple environment. Also, since we buy most of the items described, we must be alert for another reason: The sales manager, in an otherwise well-written brochure, may have made sure not to give away any competitive (or other) disadvantage of a device. This statement does not imply crookedness. Rather, we accept the fact of life that, in selling and advertising, obvious flaws are seldom stressed. Rather, they stress that their devices meet military standards which are--it is presumed--tough standards.

Thus, in each of the remaining chapters, we shall

1. Identify and correct the key misconception(s) that lead to the statement "it does not work." Usually, it is only one (or two) basic error or omission which accounts for the difficulty. In most cases the engineer will have to admit: "My analysis or

information is incomplete." And only in a few cases can the engineer blame misleading asinine or imperfect standards.
2. In some chapters, provide a brief classificatory survey of the options at our disposal. If deemed necessary, some annotations will be made. A good engineer is flexible. He or she says: "If this does not work, that one will."
3. Naturally, we shall continue to illustrate by example. And we learn even better by working problems such as those presented following Chapter 12.

REFERENCES

1. Grounding and Safety, D. J. Hatch and M. B. Raber, IEEE Transactions on Biomedical Engineering, January, 1975, pp. 62-65.
2. A Review of Lightning Protection and Grounding Practices. G. W. Walsh, IEEE Transactions on Industry Applications, March/April, 1973, pp. 133-149.
3. Trends and Practices in Grounding and Ground Fault Protection Using Static Devices, R. O. D. Whitt, IEEE Transactions on Industry Applications, March/April, 1973, pp. 149-157.
4. Neutral Currents, W. F. Bennett, IEEE Spectrum, July, 1977, p. 13.
5. Important Functions Performed by an Effective Equipment Grounding System, R. H. Kaufmann, IEEE Transactions on Industry and General Applications, November/December, 1970, pp. 545-552.
6. Protection of Structures Against Lightning, R. H. Golde, Proceedings of the IEE, Issue 10, 1968, p. 1523.
7. Electrical Transients in Power Systems, A. Greenwood, Wiley-Interscience, New York, 1971.
8. Switchgear and Control Handbook, R. W. Smeaton, McGraw-Hill, New York, 1977.
9. Optical Transmission of Voice and Data, I. Jacobs and S. E. Miller, IEEE Spectrum, February, 1977, pp. 33-41.
10. Memory Error Control: Beyond Pority, R. T. Chien, IEEE Spectrum, July, 1973, pp. 18-23.
11. Alternatives in Digital Communication, M. P. Ristenbott, Proceedings of the IEEE, June, 1973, pp. 703-721.
12. Data Transmission--The Art of Moving Information, R. J. James, IEEE Spectrum, January, 1965, pp. 65-83.

13. Applications of the Convective Null-Heat Balance Concept for Non-contact Heat Measurements, B. Birnbaum, IEEE-IAS 1977 Annual Meeting, Los Angeles, February 7, 1977.
14. Frequency Analogie, D. Gossel, ET2-A, Vol. 93 (1972), Vol. 10 (in German), pp. 577-581.
15. Use of V/F Converters, J. Graeme, Electronic Design, April, 1975, pp. 110-114.
16. Non-linear Low Pass Filter Rejects Impulse Signals, B. Gilbert, Electronics, April 1, 1976, pp. 82-83.
17. ITEM (Interference Technology Engineering Master), R&B Enterprises, P. O. Box 328, Plymouth Meeting, Pa. 19462. (Mostly advertisement; most articles by advertisers.)
18. The National Electrical Code, latest edition.
19. Recommended Practice for Commercial Power Systems Grounding Industrial (IEEE Green Book).
20. Recommended Practice for Electric Power Systems in Commercial Buildings (IEEE Grey Book).
21. Recommended Practice for Protection and Coordination of Industrial and Commercial Power Systems (IEEE Buff Book).
22. Recommended Practice for Emergency and Standby Power Systems (IEEE Orange Book).
23. Recommended Practice for Electrical Power Distribution for Industrial Plants (IEEE Red Book).
24. ANSI Standard C39.5-1975, Safety Requirements for Electrical and Electronic Measuring and Controlling Instrumentation (SAMA and NEMA).
25. Getting Noise Immunity in Industrial Controls, H. M. Schlicke and O. J. Struger, IEEE Spectrum, June, 1977, pp. 30-35.
26. IEEE Guide for the Installation of Electrical Equipment to Minimize Noise Inputs to Controllers from External Sources, IEEE Standard 518-82 (IEEE-IAS, G. Younkin, Subcommittee Chairman). An excellent companion to this book.
27. IEEE Guide for Harmonic Control, Reactive Compensation of Static Power Counters, IEEE-IAS (R. P. Stratford, Subcommittee Chairman). IEEE Standard 519-81.
28. EMC Between Electronic Traction Systems and Signaling and Telecommunications at Swedish Railways, S. Swensson, Record of the 1973 International IEEE EMC Symposium.
29. Railroads and EMC, W. H. Siemens, Record of the 1973 International IEEE EMC Symposium.
30. Interference to Telecommunication Installations by Rapid Transit Systems, R. Buckel and H. A. Riedel, Record of the 1973 International IEEE EMC Symposium.

References

31. EMC Between Thyrister-Chopper Controlled Cars and Electric Fixed Installations in Tokyo Subway, R. H. Yukawa, Record of the 1973 International IEEE EMC Symposium.

32. EMC Design for Rapid Transit Systems, I. R. Barpal, Record of the 1973 International IEEE EMC Symposium.

33. The Results of Data Transmission Impairment Due to Thyrister Control of Locomotives, P. C. DasGupta, Record of the 1976 International IEEE EMC Symposium.

34. Efforts of the SAE, IEEE, and CISPR to Control Radio Spectrum Pollution from Motor Vehicles, F. Bauer, Record of 1973 International IEEE EMC Symposium.

35. Automotive Electronics--Choosing the Right Technology, R. MacMillan, EDN, January 5, 1974.

36. What Electronic Devices Face in the Automotive Environment, O. T. McCarter, EDN, January 5, 1974.

37. Medical Device Laws: A Critical Balancing Act, R. K. Jurgen, IEEE Spectrum, July, 1973.

38. Patient Safety, Hewlett-Packard Application Note, AN 718, 1971.

39. EMC Standard for Medical Devices, MDS-201-0004, in preparation by R. J. Hoff, McDonnell Douglas Aircraft Astronautics Company East.

6 Simple Supression
What to Put Where

To get rid of interference, we distinguished between *nullifying* U* (the unwanted) and *canceling* U* - (-U*) on the one hand, and, on the other hand, *differentiating* somehow between W (wanted) and U such that $\Delta W(U^*) = 0$ (the wanted is not critically affected by the unwanted). We discussed differentiating in Section 5.3.3 because it must be considered as part of the system. Here, in contrast, we look at nullifying U*, U* = 0, and nullifying (canceling) the effect of U*, namely U* - (-U*) = 0. They are essentially localized additions, mostly not detrimental to the system performance. We shall see, however, that even very simple measures can sometimes not be considered out of context. For instance, adding an RC snubber to a contactor may slow it down unacceptably. (Why?)

Both nullifying and canceling can be done at the source S, the transfer T, and/or the receptor R. Economic considerations decide which choice to make.

In view of the ample literature available, we confine ourselves to a few very brief comments and sketches.

6.1 SUPPRESSION AT SOURCE

Resonance damping: lower Z_0 and Q (see Sec. 1.1).
Limiting d/dt [snubbing, Table 6.1(a1), (a2), (a3)] [1]; see the exercise at the end of the chapter.
Transformer isolation [2].
Harmonic suppression [3].
Switching at the zero-crossing point (ohmic load).
Switching off or limiting V or I (Table 6.2) [4,5].
Isolation trafo (common mode).

6.2 SUPPRESSION OF TRANSFER

6.2.1 Normal Mode

Twisted pairs.
Shielded cables.
Fiber optics.
Switching and limiting protection (Table 6.2) for voltage V and current I.
Staggering protective devices [Table 6.1(b2)]: a gas-filled spark gap takes the brunt; further reduction by a power zener, but since it acts faster, insert a delay line or L, operating with the C of the zener.

6.2.2 Common Mode

Balun for wire pairs; quite often ferrite shells are used for bundles of wire; steel necessary for high currents; HF only.
Isolation amplifiers (see manufacturers' catalogs).
Guarding: already discussed (see Section 1.3.2.2).
Electrooptical isolator or fiber optics.
Three-electrode spark gap to prevent common mode forming.
Balanced lines (imbalance leads to common mode).
Active compensation [Table 6.1(b1)].
Leakage current (ground-fault) switch-off.

Table 6.1 Localized Suppression

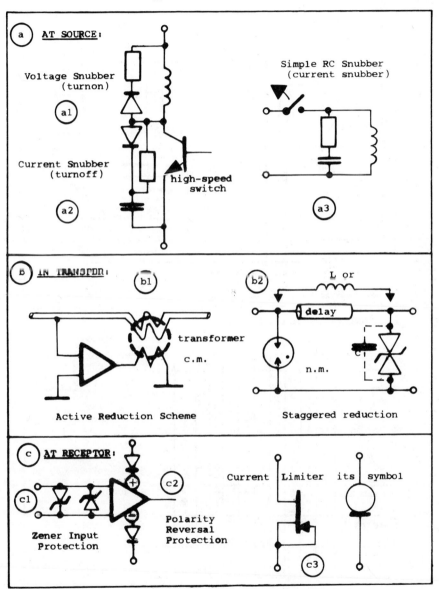

174   6. *Simple Suppression*

Table 6.2 Protective Switching and Limiting

	SWITCHING		LIMITING	
	ideal	real	nearly ideal	crude thermal approx.
V too high	(circuit diagram)	Crowbar * Spark Gap * Thyratron	MOV Power Zener	Thermistor
I too high	(circuit diagram)	Fuse Circuit Breaker	Current-Limiting Diode	Incandescent Lamp
SPEED	moderate		fast	very slow

* with limiting R or automatic turnoff

(a) broad survey

(b) important comparisons

criterion	F	A			T	T	M	E	S	S	$
	minimal C	linearity (modul.)	high voltage hdlg.	low voltage handl.	fast response	temp. insensitive	DC current stops	energy handling	insulation resist.	no aging	low price
Spark Gap (gas)	+ +	+ +	+ +			+ +	3)	+ +	+ +	+ +	+ +
MOV (GE)	+		+ +	+ +		+ +	+			+ +	
Power Zener	1)	+		+ + +	+ +		+ +	+ +	+ 2)	+ +	

REMARKS:

1) add series diode to reduce C !

2) when combining spark gap and Tranzorb use arrangment of Table 6.1 (b2)

3) for balanced lines , use three-electrode gap to prevent c.m. formation.

6.3 SUPPRESSION AT RECEPTOR

Guarding (common mode).

Differential amplifier (common mode).

Limiting V; (zeners, MOVs) (Table 6.2) [4,5].

Current-limiting diodes [6]. For instance, IN5283 limits current to 0.2 mA; important in biomedical work [Table 6.1(c3)].

Diodes for protection against polarity reversal [Table 6.1(c2)].

Exercise

In an RC circuit with a given current I_{max}

$$(dV/dt)_{max} = I_{max}/C$$

In an LR circuit with a given voltage V_{max}

$$(dI/dt)_{max} = V_{max}/L$$

Check the dimensions (V/I = R). How does τ fit in?

REFERENCES

1. Methods for Utilizing High-Speed Switching Transistors in High Energy Switching Environments, W. R. Skanadore, General Semiconductor Industries, Inc., Tempe, Ariz., 1977 (a very good paper).
2. SCR Drives--AC Line Disturbance, Isolation, and Short Circuit Protection, E. M. Stacey and P. V. Selchau-Hansen, IEEE Transactions on Industrial Applications, January/February, 1974.
3. IEEE Guide for Harmonic Control and Reactive Compensation of Static Power Converters, R. P. Stratford, Subcommittee Chairman of IEEE-IAS Subcommittee on Harmonics and Reactive Compensation.
4. Transient Voltage Suppression Manual (D. C. Kay, ed.), General Electric, 1978.
5. A Comparison Report on Transzorbs Versus Metal Oxide Varistors, General Semiconductor Industries, Inc., Tempe, Ariz., 1975.
6. The Use of Constant Current Diodes in Preventing Electrical Shock from Hospital Equipment, R. M. Fish, IEE Transactions on Biomedical Engineering, October, 1970.

7 About Shielding
The Importance of Size and Structure

This is an attempt to extract the practically useful essence from the ample, but not always satisfactory literature on shielding. For perspicuity and brevity, we make heavy use of normalization; that means introducing parameter ratios: for example, shield thickness/skin depth instead of frequency. For a better approximation of reality, we discard the plane wave approach.

The purpose of shielding is to suppress radiated interference. More than a sheet of metal is involved. For good shielding, three essential conditions must be met:

1. *Filters for all penetrating wires*. In most practical cases (except if a completely insulated, self-sufficient system is involved), a shielding box has wires going through the shield in order to transport information or energy. Thus it is imperative to filter *all* such wires which pass interference into the shielded space, however perfect and excellent the shielding may be. Thus shielding without filtering is senseless (for details, see Chapter 8). The converse is also true.
2. *Prevention of imperfections*. Other consequential leaks or discontinuities may reduce or nullify the shielding effect.

They are imperfect joints, doors, openings for ventilation, saturation effects, resonances, sharp corners, and so on. These are critical limitations, often completely unsuspected by the uninitiated. We addressed some of these problems in Chapter 3 (transfer impedance affected by holes; need for waveguides on holes; resonances).

3. *Appropriate calculations.* Finally, there is a third principal concern: how to calculate a shield. As we say in "Waves Generated Inside a Sphere (High Frequencies)" in Section 2.2.1.2 (radial impedance concept), it is convenient, but not always correct--in fact, one misses great opportunities of cost-effectiveness--if one disregards the geometry and size of the shielding structure (as is often done with the transmission line approach premised on a plane wave impinging on large plane sheets).

We shall select spheres (and in some cases cylinders) as the geometry constituting the boundary conditions when applying Maxwell's equations. In other words, a shielded room in the form of a cube will be approximated by a sphere of about the same dimensions (remember OOM). This permits us to take conveniently into account the field (originally assumed uniform) distortion by the shield, provided that we take one precaution: The box must not have sharp corners and edges, as they greatly reduce shielding--an effect that manufacturers of shielded rooms often disregard.

If, as is practically true in most cases, the thickness d of the shield is much smaller than the radius of the equivalent sphere, the solutions are rather simple, containing only hyperbolic functions instead of spherical (or Bessel) functions. Let us assume that we have gone through the rigorous process of converting the original, nonspecific partial differential equations of Maxwell, by introducing the boundary conditions, into an ordinary differential equation and have solved them [1,2,3] for the magnetic field. (See Table 7.1 as the first and rather fundamental example.) This, in a nutshell, is the result: At high frequencies, the electric

Frequency Domain

Table 7.1 Magnetic Shielding by Metallic Spheres

TERMS :
$\mu_o = 4 \cdot 10^{-7}$ H/m
μ_r = relative permeability
σ = specific conductivity
δ = equivalent skin depth
MKS(r) system

CONDITIONS OF VALIDITY:
$d \ll r_o$
no leaks
no resonances
external excitation

SOLUTION TO MAXWELL'S EQUATIONS:

$$\ln a_S = \cosh \bar{k}d + (1/3)(K + 2/K)\sinh \bar{k}d \tag{7.1}$$

where
$$K = \bar{k}r_o/u_r \tag{7.2}$$
$$\bar{k} = (1 + j)/\delta \tag{7.3}$$
$$\delta = (2/\omega \mu_o \mu_r)^{0.5} \tag{7.4}$$

or in explicit form with
$$q = r_o/u_r \delta = (r_o/u_r d)(d/\delta) \tag{7.5}$$
$$v = 2d/\delta \tag{7.6}$$

$$a_S = (1/2) \ln \Big\{ (1/9)(q^2 + q^{-2})(\cosh v - \cos v) \\ + (q/3)(\sinh v - \sin v) \\ + (1/3q)(\sinh v + \sin v) \\ + (1/2)(\cosh v + \cos v) \Big\} \tag{7.7}$$

SPECIAL CONDITIONS :

(a) <u>Magnetic material, very low frequencies</u>: $q \ll 1; \delta \to \infty; u_r \gg 1$
(use Eq.(7.7)) with $\bar{k} \to 0$)
$$a_S \to \ln\left[1 + (2/3)(u_r d/r_o)\right] \tag{7.8}$$

(b) <u>Non-magnetic material, low frequencies</u>: q 1; d ; $u_r=1$
$$a_S \to (1/2) \ln\left[1 + (2/3)(r_o/\delta)(d/\delta)\right] \tag{7.9}$$
(for $f = 0$, $\delta = \infty$ and $a_S = 0$)

(c) <u>Magnetic & non-magnetic (q 1) materials</u>: $d > \delta$
$$a_S \approx (d/\delta) + \ln(r_o/3\sqrt{2}\,\mu_r \delta) \tag{7.10}$$
$$\approx (d/\delta) + \ln 1/3\sqrt{2}(r_o/\mu_r d)(d/\delta) \tag{7.11}$$

In all other cases use Eq. (7.7), in which $1/q^2$ and the $1/q$ term should be neglected for the non-magnetic case.

---

and magnetic fields are equally well shielded by good continuous shields, but for very low frequencies, the magnetic field is much more difficult to shield (no or very low eddy currents). The size of the shielding enclosure plays a significant role, having opposite effects for magnetic and for nonmagnetic materials at low frequencies. We shall investigate this in more detail and will learn how to cope with the detriments imparted by actual structures.

A special section (7.2) is arranged for shielding pulses. This is a rather ticklish problem: Shape factors, frequency dependency over broad ranges, and nonlinearity interact complexly. So we have to be cautious in selecting an appropriate simplifying model that yeilds a realistic, useful answer, not just a mathematical exercise. The time needed to penetrate a shield turns out to be a key criterion. For ferromagnetic materials, the details of the magnetization curve become secondary. Rather, H_c, the coercive force, and B_s, the saturation density, suffice to estimate shielding penetration in the time domain.

7.1 FREQUENCY DOMAIN

7.1.1 Metallic Shells

7.1.1.1 Single Solid Shells (Table 7.1, Figs. 7.1 and 7.2)

We establish a universal relationship by introducing d/δ, the ratio of the thickness of the shield and the equivalent skin depth (a function of frequency and material "constants"), as a parameter (Eq. 7.4). d/δ is also a direct measure, in nepers, of the absorption loss of the shield

As abscissa, we take the ratio r_0/d, the radius of the equivalent sphere divided by the thickness of the shield. However, this factor must be multiplied by $1/\mu_r$, where μ_r is the relative permeability of the shielding material. Hence there is a great difference in our scale factor, depending on whether or not the shielding material is magnetic. Figure 7.1 permits us to read the equivalent skin depth. With it, we can enter the universal Figure 7.2, which is based on the general equation 7.7 of Table 7.1 (representing the solution to Maxwell's equations for spheres). If we work with nonmagnetic material, we will be operating on the right side of Figure

Frequency Domain 181

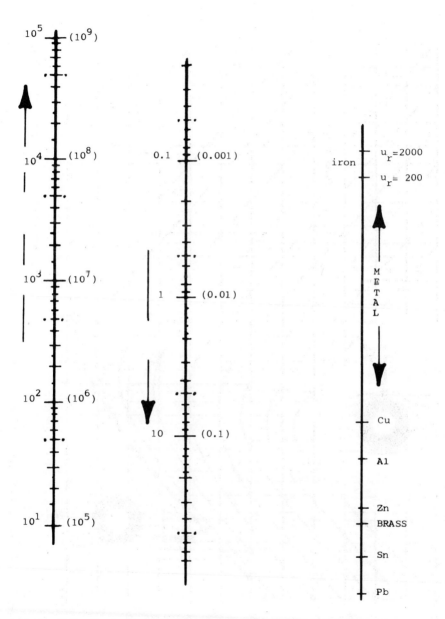

Figure 7.1 Equivalent Skin Depth

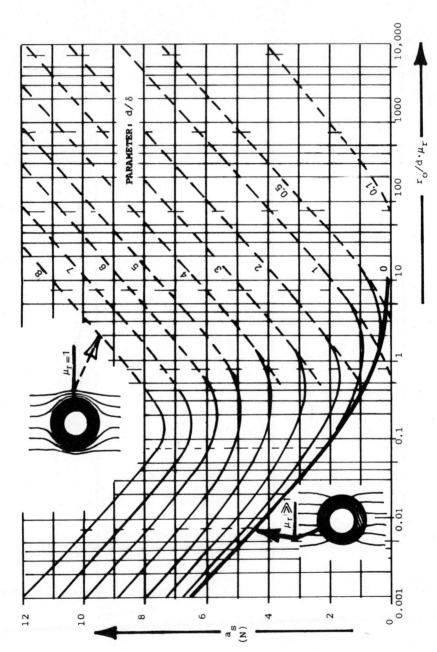

Figure 7.2 Shielding by Magnetic (Left) and Nonmagnetic (Right) Spheres

7.2 ($r_0/\mu_r d \ll r_0/d$, or the $1/q^2$ and the whole term containing $1/3q$ are discarded). For a definition of δ, see Table 7.1. For the nonmagnetic case, these curves go monotonously to 0 for lower q's, meaning that dc magnetic fields, for lack of eddy currents, are not shielded by nonmagnetic metals. This is indicated in Figure 7.2; the curves for nonmagnetic materials do not bend upward for lower q, but tend to 0, as hinted with the dashed lines at the low a_s values of d/δ = 4, 3, 2, 1, ...

In complete contrast, for magnetic materials we work with q < 1 (because $\mu_r \gg 1$); then the term $1/q^2$ in Equation 7.7 is dominant. Automatically, we now operate on the left part of Figure 7.2 ($r_0/\mu_r d < 1$). Surprisingly, we find better shielding for $r_0/\mu_r d$ becoming smaller. In fact, even for $(d/\delta) \to 0$, that is, very low frequencies, smaller geometries become better shields (whereas in the nonmagnetic case, the opposite is true). Physically, this can be explained by the distortion of the (originally assumed uniform) magnetic field by the shielding structure. In the $\mu_r = 1$ case, the field lines flow around the structure. In the magnetic case, for lower frequencies, as Figure 7.2 also shows, the field lines are sucked into the structure (for high-frequency fields $\delta \to 0$ and the magnetic shells have a field configuration like the nonmagnetic shells).

Here we encounter a very peculiar and practically very significant situation that merits examplification.

Example 7.1 Size Effect--Magnetic Case. How does a change in radius (a) r_0 = 1 cm; (b) r_0 = 100 cm, affect the shielding for the following conditions: μ_r = 2000; $d/\delta < 1$; d = 1 mm? For (a), $r_0/\mu_r d$ = 0.005. The shielding a_s is 4.8 Np. For (b) $r_0/\mu_r d$ = 0.5 or a_s = 0.8 Np. That is, for low frequencies, the magnetic shield given has a shielding effect of 4.8 Np for the smaller sphere or 0.8 Np for the larger sphere. At high frequencies, the shielding is improved by absorption losses. According to Figure 7.1, δ = 0.2 mm at 2 kHz for iron with μ_r = 2000. Hence with d/δ = 5, Figure 7.2 yields 7.3 and 5.0 Np, respectively, for the small and larger spheres at 2 kHz.

Example 7.2 Size Effect--Electric Case. Now let the same two spheres be made of copper. At dc the magnetic field is not at all shielded. The skin depth for copper is 0.2 mm at 100 kHz → d/δ = 5 (Figure 7.1). For (a) r_0/d = 10 and (b) r_0/d = 1000, we find (a) 7.5 Np and (b) 12 Np.

The juxtaposition of Examples 7.1 and 7.2 shows clearly the advantage of small boxes for magnetic shields and large boxes for unmagnetic shields. At 2 kHz, copper has a skin depth of 1.5 mm, that is, d/δ = 0.66. The small copper sphere would shield only 1 Np, the large one 5.5 Np.

What is it we learn? The geometry plays a significant role for relatively thin boxes, and magnetic boxes have opposite dependencies on the box dimensions.

There is another highly important advantage of choosing the smallest possible shell to shield against static or quasistatic magnetic fields: It has to do with saturation effects. When the magnetic field is sucked into the shielding wall, the field is enhanced by a factor of 3, independent of size (see Sec. 2.1.2.2). That is, $3H_0$ is the effective field that flows through the area $2\pi r_0 \cdot d$, resulting in a flux density of $3H_0 \mu_0 \mu_r$ collected by an area of $r_0^2 \cdot \pi$. We assume that $d < \delta$.

Example 7.3 Bias Effect. How does this field concentration affect the biasing of the magnetic shell (localized at the equator)? Assume that H_0 = 0.2 Oe. Since many magnetic data are still given in the cgs system, we make use of the magnetic conversion factors given in the following table:

MKS (r)	CGS System
B (Wb/m^2)	10^4 B (G)
H (A/m)	$4\pi \cdot 10^{-3}$ H (Oe)
μ_0 (H/m)	$10^4 (4\pi)^{-1}$ BH^{-1} (G/Oe)

$3H_0 = 3 \times 0.2$ Oe = 0.6 Oe or 0.075 H/m, or for air a flux of 0.075 Wb/m^2. This flux is now concentrated by $\pi r_0^2/2\pi r_0 d$ or $r_0/2d$, that is, for (a) 10 times; for (b) 1000 times, or the flux density at the equator of the sphere is 0.75 Wb/m^2 for (a) and 75 Wb/m^2 for (b). This means that the material is completely saturated in case (b) and one might as well have used wood or cardboard for shielding. In contrast, the smaller radius shows only moderate bias effects.

For magnetic cylinders, the 3's in Equation 7.7 have to be replaced by 2's, resulting in a diagram such as that of Figure 7.3. Some other, more specific conditions, are presented in Figure 7.4.

7.1.1.2 Chicken Wire

Based on calculations made by Kaden [2], Figure 7.5 allows the determination of wire grid shields (they can also be perpendicular gratings without interconnection). Such chicken wire screens are suitable for shielding at higher frequencies. Since they operate on eddy currents, the shield must be well soldered to permit passage of circulating currents around the cage. Structures of steel-reinforced concrete are rather effective shields *if* the flow of eddy currents is permitted by proper welding of the steel rods (see also Problem 74).

Example 7.4 Wire Mesh. A printing plant, using minicomputers, is located in a building that also houses plastic welding machines, working at about 30 MHz, causing annoying interference in the printing plant. One of the options for interference control is the use of chicken-wire-shielded rooms. Suppose that we use chicken wire, zinc plated, with a wire radius of r_w of 0.5 mm and mesh width s of 1 cm, and that the smallest dimension of the room is $2x_0$ or 5 m. With $r_w/s = 0.5/10 = 0.05$ and $x_0/s = 250$, we find an asymptotic attenuation of 7 Np which will occur if the interfering frequency is well above f_c, as defined in Figure 7.5. From Figure 7.1, we may find $\delta_w = 0.02$ mm. Thus $r_w/\delta = 0.5/0.02$ or 25, which is well above 1.4.

Naturally, we can expect reduced shielding at very high frequencies where the openings of the screen are on the order of $\lambda/2$.

7. About Shielding

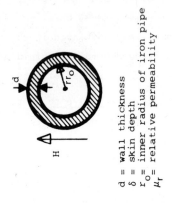

d = wall thickness
δ = skin depth
r_o = inner radius of iron pipe
μ_r = relative permeability

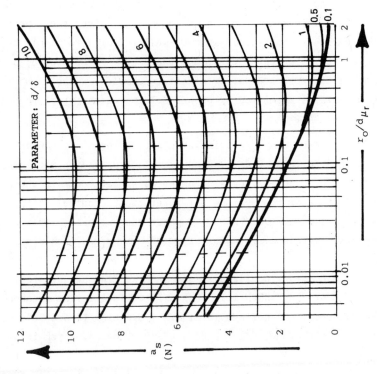

Figure 7.3 Shielding by Magnetic Cylinders

Frequency Domain

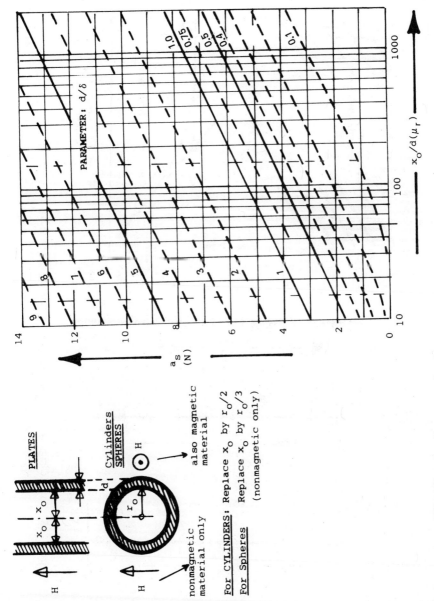

Figure 7.4 Shielding by Nonmagnetic Structures

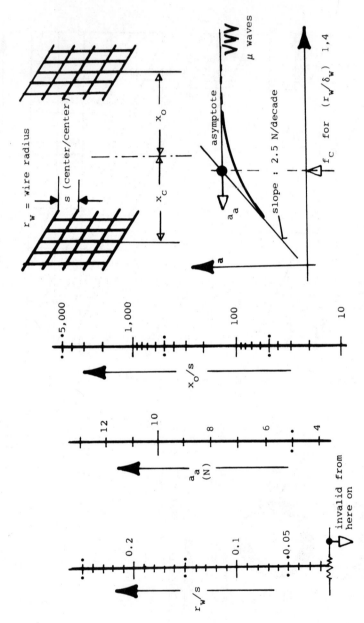

Figure 7.5 Shielding by Wire Grids (Disregarding VHF "Holes")

Frequency Domain

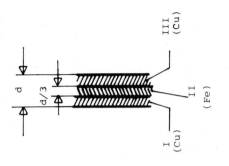

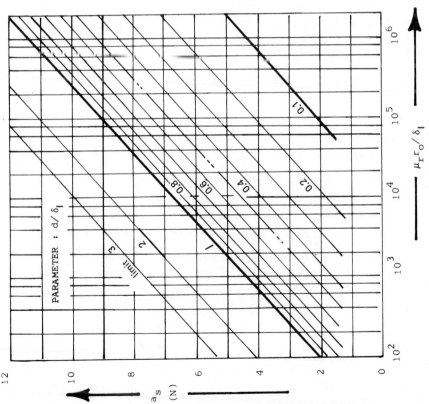

Figure 7.6 Optimal Three-Layer Shield

7.1.1.3 Laminations

Also based on calculations by Kaden [2], Figure 7.6 shows what shielding can be achieved by optimized lamination (as indicated) for low frequencies (kHz range). Note the different abscissa and the fact that the straight-line presentation is limited to the ranges plotted.

7.1.1.4 Coping with Reality

We now want to elaborate a bit on the imperfections mentioned at the beginning of the chapter.

1. *Permanent joints.* These should preferably be welded. If riveted, the rivets should be very close and so tight that corrosion and oxidation cannot penetrate the originally clean metallic interface to be joined.
2. *Not permanent joints.* For doors of shielded rooms, we use double rows of finger stock of noncorrosive spring material. For assemblies, one should contact manufacturers of resilient seam material. Check for loss of resiliency (aging) under compression. Confine the seam material. For large doors, hydraulic compression of seams is excellent. Prevent oxidation of the seam material and of the material to be joined. Avoid silver in atmospheres containing sulfur.
3. *Holes.* See Sections 2.1.1.3 and 2.1.1.4.
4. *Edges and corners.* It is beyond the scope of this text to discuss in any detail the "edge effect" of shielded rooms [2, pp. 91-107]. The important fact is that even for nonsaturating materials, the field in corners and edges is greatly enhanced, the more so the closer one comes to the corner. Rounding of corners and edges or at least no measurements close to them is strongly advised.
5. *Resonances (external and internal excitation).* Dips up to 20 or 30 dB happen in the high-frequency, high-attenuation range, where the dimensions are on the order of $\lambda/2$ or larger (cavity) effects, microwave effects shifted to VHF, UHF frequencies; see Ref. [2]).

6. *Resonances (internal excitation, standing waves).* Measurements using transmitters in the shielded room are severely falsified by standing waves (same cavity effect as above), as the measurement may be the response to the minimum or maximum of a standing wave. The following remedies are available:
   A. Apply forests of lossy microwave absorber cones inside the shielded room. Danger: They may ignite under high-power excitation. Also, this treatment is expensive.
   B. Mode stirred chambers, using moving reflectors in the shielded room to shift the minima periodically.
   C. Holes in the earth, with a transition layer (see the second subsection of Section 2.2.1.2 for detailed discussion and derivation).
   D. The most radical approach: Shift from frequency domain to time domain by replacing a continuous-wave (CW) frequency transmitted by a pulsed dc transmitter feeding a Crawford cell.

7.1.2 Nonmetallic Structures

See the first subsection of Section 2.2.1.2 for the quasistatic cases of μ' or ε' shields only. See the second subsection for the high-frequency absorption of soil.

7.2 TIME DOMAIN

7.2.1 dB/dt Reduction by Shields

See Section 2.2.2, where we exemplified the drastic dB/dt reduction of pulses for plane wave, linear conditions (low-pass filter analogy).

7.2.2 Pulse Transfer in Cables

In Tables 7.2 (nonmagnetic case) and 7.3 (magnetic case), we pro vide two handy nomographs constructed on the basis of the excellent reference [5]. These nomographs contain all the information needed for their intelligent use.

Table 7.2 Impulse Shielding (Nonmagnetic Case)

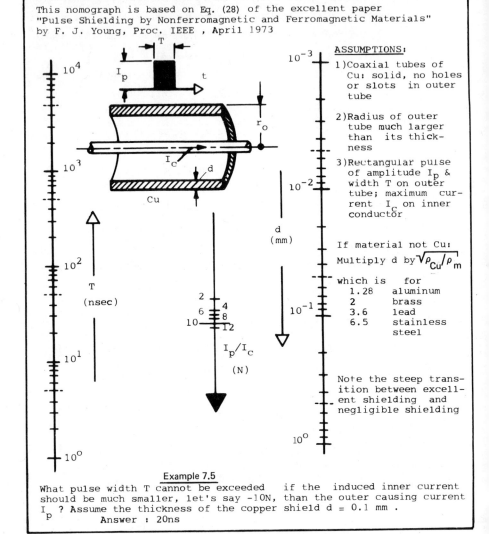

Table 7.3 Impulse Shielding (Magnetic Case)

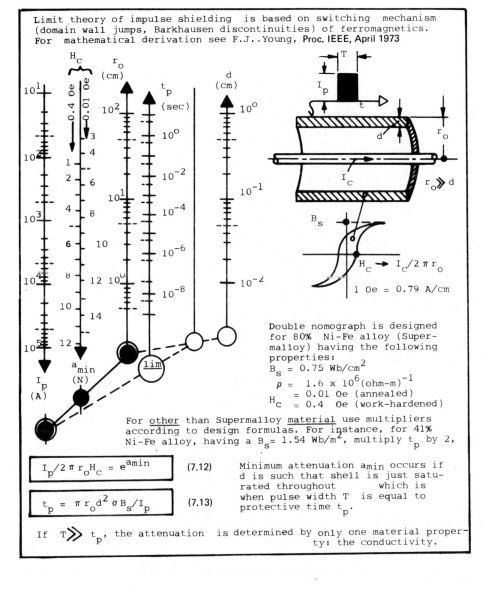

7.3 CONCLUSION

This was not an easy topic which we tried to make digestible and manageable in very concentrated form. Nevertheless, it is hoped that the reader gained the perspective and tools indispensible for a reasonable and practicable handling of shielding problems. They encompass all eight FATTMESS criteria and often do not stand constraining simplifications, which may cause great errors.

REFERENCES

1. IEEE Transactions on EMC, Special Issue on Shielding, R. B. Schulz, editor, March, 1968.
2. Wirbelstroeme und Schirmung in der Nachrichtentechnik, H. Kaden, 2nd ed., Springer-Verlag, Berlin, 1959 (in German).
3. Electromagnetic Waves, S. A. Schelkunoff, Van Nostrand, New York, 1945.
4. Potentialfelder der Elektrotechnik, F. Ollendorf, Springer-Verlag, Berlin, 1932 (in German).
5. Pulse Shielding by Non-ferrous and Ferromagnetic Materials, F. T. Young, Proceedings of the IEEE, April, 1973, pp. 404-412.

8 Filtering for EMC
Throw Away Your Filter Books

The purpose of filtering is to suppress conducted interference. The reader may ask: Why a chapter on filtering when there are already so many filter books on the market? The answer is simple: Conventional filter books treat filters for highly idealized and simplified conditions that are not valid for many an EMC situation, as we discussed briefly in Section 1.3. And because of the importance of this speciously simple situation, we have to dig deeper.

Quite often when the engineer unthinkingly applies what the book says, he expects that the filter works by a miracle in the nonideal and complex EMC environment. But unfortunately, miracles do not happen so often. It is a fact that filters often do not do what they are supposed to do. A rather crass example happened recently. A filter manufacturer had supplied the filters for a large system. The filters met all standards (military). Still the system behaved erratically because of filter dysfunction (ringing and insertion gain). Thus the company had to file for bankruptcy. In many a Tempest, installation filters are thrown out because of ringing.

Such difficulties are not caused by the "difference between theory and practice;" rather, the difficulties stem from

1. Choosing the wrong class of filters (FATTMESS).
2. Disregarding the operating conditions (bias, mismatch, temperature, etc.).
3. Incorrectly installing the filter (no shielding).

Instead, then, of ruminating about conventional, not always applicable filter theory, we have to generalize existing filter theories and marshal the adequacies of various filter classes for real, multiconditional EMC situations requiring careful analysis.

We have here the peculiar and unexpected situation that the oldest, the simplest, and the theoretically well founded filters, namely LC filters, cause much confusion and frustration due to misapplication. (Severe mismatch is applied to simplified theory valid only for matched conditions.) In fact, the situation, admittedly complex, is so bad that we must make a real effort at its clarification.

In Section 5.3.3 we already analyzed such conditional filter analysis/control, which is, by necessity, an integral part of systemic planning. There we considered filtering in its broadest sense, encompassing all FATTMESS criteria. We shall not repeat this here. We shall also not set forth the excellent presentations available on active, preatersonic, correlation, and other types of filters.

8.1 THE INSIDIOUS PROBLEM

Classical, passive filter theory, based on early fundamental work by Campbell and Kauer, is well developed for communication systems where one can relatively easily operate under impedance-matched conditions ($Q = 1$). In communication systems, impedance matching is stipulated and held within rather narrow limits. If such conditions are attainable in EMC work, for instance if unidirectional buffer stages can be inserted without great cost, no serious filter problems arise and one can follow standard filter books.

Assumptions of matching are, however, very contrary to fact in power feed lines on which loads are continually switched. To treat such a situation as if there were impedance matched or to take the statistical mean is convenient and speciously scientific but definitely wrong.

Power feed lines permeate (and spread the noise ubiquitously throughout) the whole system and neighboring systems they feed. Such power feed lines and their generators and loads are designed for the singular purpose of high efficiency at 60 Hz or whatever the frequency is. Moreover, the heavy current in the line may saturate the inductors of the filter, thus making the inductor inoperative. Altogether, then, filtering for EMC is just not conventional filtering. And assuming that both kinds of filtering are identical is the same as declaring cough medicine and medicine for diarrhea to be identical because both are medicines. Consequently, we shall have to attack the problem of EMC filters without reliance on established (but not applicable) filter theory and filter measuring techniques to make sure that the filter will work in the real environment, not only in the textbook world. And the reader will probably be surprised that we put so much emphasis on passband behavior, which in conventional filter theory is rarely suspected of causing any trouble.

8.2 THE SOLUTION FOR INDETERMINATE MISMATCH

8.2.1 Plan of Attack

So far, there has been no satisfactory solution to the perplexing, dual problem that filters often do resonate in the passband or do not attenuate in the stopband as they are expected to do. It is true that many well-meant approaches to the problem exist, each treating a partial, simplified aspect that does not constitute the whole problem pertaining to time-variant mismatch effects, let alone the complicating influence of bias. The heretofore unsatisfactory state of our capability to cope with the mismatch problem is clearly indicated by the absence of meaningful and satisfactory methods applicable for measuring the insertion loss (I.L.) of LC filters. Many such methods exist, none being adequate. Measurement of LC filters under *any* worst-case conditions (that are unrealistic) or measuring them in situ (which is impractical and does not predict future operating) does not provide satisfactory solutions. Worse "solutions" are those which are premised on an average (mostly rather high) impedance for generators and loads (e.g., Mil.Std.220A, measuring the filter inserted into a system where the generator and loads are "approximated" by 50 ohms, treats as we shall see, a very uncritical condition). Similarly misleading, CISPR measurements are based on absorption clamps, constituting relatively high resistive loads at high frequencies. And line stabilization networks, required according to other military specifications, are employed for measurements, but not in the real filter operation. (This is like cutting off the branch on which one sits.)

Altogether, such methods based on statistical averages are about as good as the conclusion of the mathematician who could not swim but waded confidently into the river because he had calculated the average depth of it to be 0.5 m and concluded that the river was therefore safe. When he applied his conclusion, his theory was shaken only for a short time because he drowned rather rapidly.

To get a handle on what seems to be completely unpredictable, we shall proceed as follows:

1. Establish, empirically, a statistical data base on only the real interface impedances, not *any* imaginable ones. Such a delimiting to reality will exclude many realms of impedances that would necessitate large and expensive filters.
2. Try, empirically, to subject various filter configurations of different characteristic impedance Z_0 to all reasonable combinations of interfacial impedances thus established. This can be done vicariously rather economically on a computer (instead of actual measurements), hopefully to establish meaningful criteria and to help in the categorization of the interfacial impedances.

The Solution for Indeterminate Mismatch

3. Develop a meaningful and predictive theory of filter behavior imparted by the interface impedances encountered in the real world (but we shall not make the mistake of basing our theory on averages, as the mathematician did).
4. Apply this filter theory and the realistic boundary conditions (Z_G, Z_L) to the design and measurement of filters that do work predictively. (G = generator; L = load)
5. Provide some other pertinent comments on practical filter design, such as bias effects and feed-through requirements.
6. Later we shall apply our findings to develop meaningful filter measurement methods (Chapter 11).

8.2.2 Data Base, Categorization, and Heuristic Approach

With the aid of a government contract (NOL), a statistical data base for the interfacial impedances of 60 Hz generators (G) and loads (L) has been established for many hundreds of outlets and devices used in households, factories, laboratories, aboard naval vessels, and so on. The results are summarized in Figure 8.1 for normal mode and in Figure 8.2 for common mode conditions.

A rather surprising fact stands out: Except for common mode loads, low Q's are highly dominant. Besides the absolute impedance value, the graphs provide the limits of the few high Q's (Q maximum). They are marked (-j) for capacitive and (+j) for inductive interface impedances. Most of the Q's are less than 2 or 3. A typical example is given in Figure 8.3, representing in more detail Figure 8.1(d), normal mode impedances and Q (±j) of regulated power supplies. As will soon become obvious, it is no serious drawback that some of the data do not extend beyond 100 kHz, where knowing them is unimportant. Some comments on specific cases are in order:

Figure 8.1(a) presents the normal mode impedance of 60 Hz outlets, unregulated. The lower, less densely shaded area pertains to very high current outlets.

Figure 8.1(b) shows the change if regulated outlets are measured. A significant difference exists in the normal mode loads:

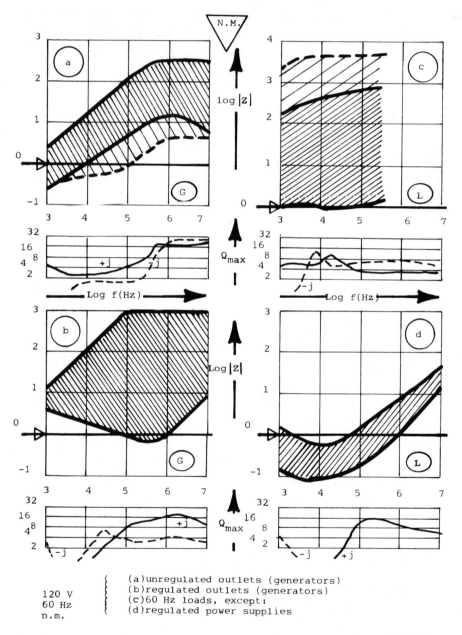

Figure 8.1 Normal Mode Interface Impedances (U.S.), 120 V, 60 Hz: (a) Unregulated Outlets (Generators); (b) Regulated Outlets (Generators); (c) 60 Hz Loads, Except as Noted; (d) Regulated Power Supplies

The Solution for Indeterminate Mismatch

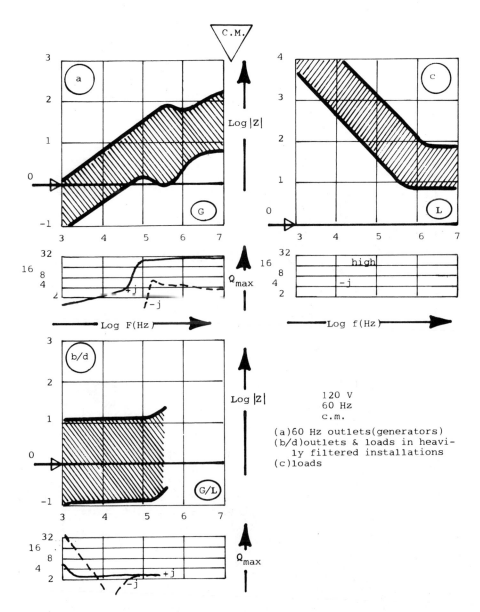

Figure 8.2 Common Mode Interface Impedances (U.S.), 120 V, 60 Hz: (a) 60 Hz Outlets (Generators); (b/d) Outlets and Loads in Heavily Filtered Installations; (c) Loads

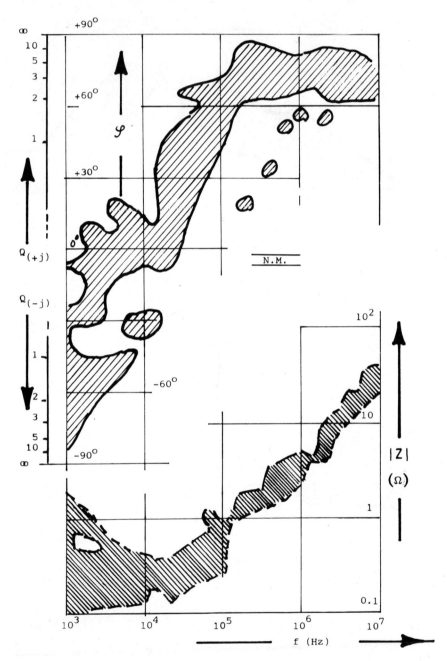

Figure 8.3 Impedances and Q of DC Power Supplies, Loaded and Unloaded

Figure 8.1(c) portrays all loads (with the top, less densely shaded area representing high-voltage, low-current power supplies; the middle range pertaining to motors, TV sets, and so on; and the low limit dominated by high-current devices). Figure 8.1(c) does not contain regulated power supplies, which are singled out in Figure 8.1(d) (see also Fig. 8.3), for they constitute a distinct class by themselves because of their significantly lower impedance limits. We shall understand the need for this distinct categorization quite clearly when we will discuss the theory of mismatch. For the moment, we indicate only why regulated power supplies have the low range of Z_L. Assuming, to start with, an efficiency of 100%, the dynamic resistance of these constant voltage devices is negative

$$R = \left(\frac{dI}{dE}\right)^{-1} = -\frac{E^2}{P}$$

where $I = P/E$ (P - power). The negative resistance is in series with the positive resistance, being equivalent to the losses of the supply having an efficiency of less than 1. Hence the regulated power supplies have a low load resistance for normal mode.

Figure 8.2, pertaining to common mode interfacial impedances, says that the impedances of unregulated power outlets are about one-third of those for the normal mode. In common mode operation, lines have smaller L, larger C. Figure 8.2(b/d) depicts generator and load impedances in heavily filtered installations, as encountered, for example, in ship installations. Again, a critical low limit of 0.1 ohm exists. Figure 8.2(c) shows the typically capacitive behavior of all sorts of loads [excepting (b/d)] measured in the common mode.

The mass of data, summarized in Figures 8.1 and 8.2, suggests the use of a computer to investigate heuristically the behavior of various filter configurations in real-world interface impedance conditions. Figure 8.4 is the flow graph employed for this purpose, with Equation 8.1 defining the insertion loss. In contrast to attenuation (which is the ratio of the output voltage to input voltage under matched conditions), the insertion loss is defined as the

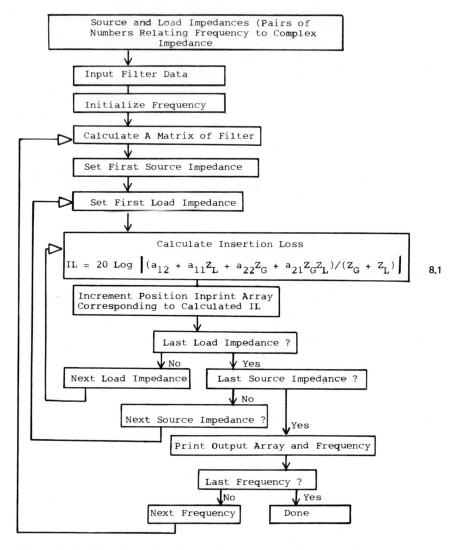

Figure 8.4 Flowchart for Insertion Loss Calculations

log of the insertion ratio of the load voltage (or current) measured without a filter inserted in the load voltage (or current), with the filter being inserted between the generator and the load. In Equation 8.1, the a_{mn} are the terms of the cascade (chain) matrix by which the filter is *uniquely* defined and the Z_G and Z_L are,

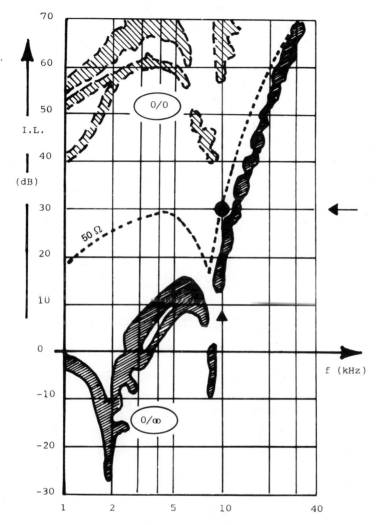

Figure 8.5 0/0 and 0/∞ versus 50/50 ohm Interfaces

respectively, the generator and load impedances that the filter faces (boundary conditions of the filter).

It is true that a computer can print out reams of insertion loss plots for all kinds of filters interfaced with all possible combinations of Z_G and Z_L. This vicarious measurement approach (quasi-in-situ) was actually done, resulting in piles of printouts. Yet such a massive attack did not yield very useful results. No practical filter was found that would have been a universally acceptable and effective filter. Tentative categorization of Z_G and Z_L (see Figs. 8.1 and 8.2) and nondimensionalization of Z_G and Z_L in terms of Z_0, the characteristic of the filter, brought about some transparency.

Example 8.1 Clues from the categorization are quite obvious in Figure 8.5, relating to a specific filter configuration designed to have 30 dB insertion loss at 10 kHz in a 50/50 ohm system (dotted line). In this context it does not matter what this particular filter configuration is, but it is important that the filter was subjected (vicariously on the computer) to two sets of real interface conditions: 0/0 and 0/∞. For simplicity, we introduce a convention that we shall use throughout this chapter. 0 denotes Z_G/Z_0 or $Z_L/Z_0 \ll 1$ (short-circuit approximation), and ∞ denotes Z_G/Z_0 or $Z_L/Z_0 \gg 1$ (open circuit approximation). The symbol in front of the slant refers to Z_G, the one following the slant to Z_L; hence 0/∞ means: low generator impedance/high load impedance, both normalized to Z_0. Whereas in Figure 8.5 the insertion loss for the 0/0 condition always exceeds the 50 ohm performance, the situation is rather dismal for the 0/∞ condition. Insertion gains (negative insertion loss) up to 26 dB occur in the passband. Thus, although the heuristic approach yielded some insight, it was abandoned in favor of a systematic analytical approach. But the empirical-heuristic approach will be reinstated to check the adequacy of the theory of mismatch filters to be developed next.

8.2.3 Theory of Indeterminately Mismatched Filters

In our effort to resolve the long-standing and heretofore unresolved controversy about the behavior of unmatched filters, we must forget textbooks on filters--they are of no avail (because one cannot draw valid conclusions from the particular to the general, but only from the general to the particular). Rather, we have to start from scratch, that is, with Equation 8.1 (Fig. 8.4). It looks like a formidable task to obtain a clear understanding of the mismatch effect expressed by this equation. But it only looks that way. Inspection of Figures 8.1 and 8.2 reveals that at the critical frequencies 1 to 100 kHz, the interfacial impedances have essentially a low Q (except common mode loads, which we can handle with bifilarly wound lossy coils). This general characteristic of our Z_G's and Z_L's precludes the existence of any sharp resonance between the filter and one or both of the interface impedances. Such interactive (interfacial) resonances will therefore be of secondary importance and can, if desired, be removed (Example 8.3, to follow). Thus our main concern will be with eigen (self) resonances of the filter, caused by open or short circuit conditions or their approximations. These are also the frequencies at which the filter rings. To begin with, we subject (Eq. 8.1) a filter characterized by the a_{mn} matrix to the extreme interface conditions $0/0$, ∞/∞, $0/\infty$, and $\infty/0$, all normal mode. The objective is to find out (1) why the filters do not always filter in the stopband, and (2) why they sometimes have pronounced eigen (self) resonances (negative insertion loss). Having found conditions for such dysfunctions [1], we will be in a position to devise corrective measures to prevent such dysfunctions in an economical way.

We have marshaled five tables to achieve this end. In Table 8.1, we ask the question: What is the most simple (primitive) and the next-higher-order filter configuration for the four possible sets of extreme interface conditions? We also indicate the reasons for insertion gain (-I.L.), although we discuss this in more detail later (Table 8.4).

For case I ($Z_G = \infty$, $Z_L = \infty$), we see, by inspection, that a shunt capacitance is the most primitive filter that is effective. For case II (0/0), it is a simple series L (or ferrite beads at high frequencies). The reader will easily continue this line of reasoning for the $0/\infty$ and $\infty/0$ cases, the general rule being: For best results, the high interface impedance must meet a C (shunt) and the low interface impedance must meet an L (series) of the filter.

Thus there is no problem if one knows the impedances the filter has to work with. But now comes the rub: In practical installations,

Table 8.1 "Best" Primitive Filters

case	Z_G	Z_L	I.L.= 20 log $\|x\|$	a_{mn}	most primitive effective filter	next, less primitive filter	danger of insertion gain (-I.L.)
I	∞	∞	$a_{21} \cdot \infty$	$j\omega C$ for π : $j\omega C \times \left[2 - \left(\frac{\omega}{\omega_0}\right)^2\right]$			NO
II	0	0	$a_{12}/0$	$j\omega L$ $j\omega L \times \left[2 - \left(\frac{\omega}{\omega_0}\right)^2\right]$ for T			NO
III	0	∞	a_{11}	$1 - \left(\frac{\omega}{\omega_0}\right)^2$			YES
IV	∞	0	a_{22}	$1 - \left(\frac{\omega}{\omega_0}\right)^2$			YES

$\omega_0 = (LC)^{-0.5}$; $Z_0 = (L/C)^{0.5}$; ∞ and 0 mean large or small in terms of Z_0

The Solution for Indeterminate Mismatch

one hardly knows the interface impedances. And even if one would measure them at an instant, they would be different at another time. Continuous switching within the system and additions and changes make the interfaces unavoidably indeterminate. We shall now, in some detail, show how to manage this indeterminacy effectively and economically. We shall do this by partitioning and damping.

Example 8.2 Partitioning. In order to prepare for the discussion to follow, let us derive, by simple matrix algebra, the next-higher-order filter for (a) the shunt capacitor (a pi filter) and for (b) the series inductor (a T filter):

CONDITION	∞ / ∞	0/0
primitive filter	shunt C	series L
next higher filter	L, C (π network)	L, C (T network)
equivalent a_{mn}	$a_{21} = j\omega C\left[2-(\omega/\omega_0)^2\right]$	$a_{12} = j\omega L\left[2-(\omega/\omega_0)^2\right]$ $\omega_0 = (LC)^{-0.5}$
for higher frequencies $\omega > \omega_0$	$a_{21} = j\omega C (\omega/\omega_0)^2$	$a_{12} = j\omega L (\omega/\omega_0)^2$

In other words, by simple partitioning of the most primitive filters, their equivalent C or L values are made to increase with the second power of frequency, for $\omega > \omega_0$. The advantages of partitioning (subdivision) filters, however, goes much further than Example 8.2 would suggest. In Table 8.2 and the accompanying Table 8.3, we have drawn up the many beneficial effects of partitioning for the 0/∞ interface, requiring for the primitive case a single L/C network and for more effective filtering, a cascading (partitioning) of LC networks (LCLCLC, etc.) (Table 8.3). The premise of partition we use is as follows: Maintain the same total capacity irrespective of how much we divide (there are, however, limits to the effectiveness of division if n becomes too large or a_c is too

Table 8.2 Partitioning of Filters

PREMISES :

1) $0/\infty$ - interface : $Z_G = 0$; $Z_L = \infty$
2) total Capacitance is constant : $C_n = C_1/n$.
3) (circular) corner frequency f_c is defined as that frequency at which attenuation is a_c (high, not 3 dB).
4) a_c is identical for all filters (fixed-C partition)

$$a_c = V_i/V_o = \omega_c^2 L_1 C_1 = \omega_c^{2n} L_n^n (C_1/n)^n \quad \text{at } \omega_c!$$

Hence $L_n/L_1 = a_c^{-(1 - n^{-1})} \cdot n$ (8.2)

$$\omega_c = a_c^{0.5} \cdot \omega_o$$

and for $n = 1$:

$$\omega_o = (L_1 C_1)^{-0.5}$$

5) relative price (conservative estimate) $RP = L_n C_n \cdot n^3$, because price is roughly proportional to L_n, C_n number of L's and C's and number of feed-through shields (of C's).

RESULTS OF PARTITION :

	n	L_n/L_1	C_n/C_1	Z_{on}/Z_{o1}	RP
see also **Table 8.3**	1	1	1	1	1
	2	0.063	0.5	0.355	0.252
	3	0.03	0.333	0.3	0.267
	4	0.0225	0.25	0.3	0.36

ADVANTAGES OF PARTITION :

1) in stopband : very much reduced mismatch.
2) in stopband : very much reduced effect of filter configuration.
3) steeper cutoff.
4) smaller size.
5) less expensive : optimum at n = 3.
6) less bias sensitive because of smaller L.
7) higher eigen resonances } to be exploited later
8) smaller Z_o

BUT PARTITION CANNOT   (a) eliminate insertion gain
  (b) ringing,

Table 8.3 Eigenresonances as Functions of Partition Number and Interface Conditions

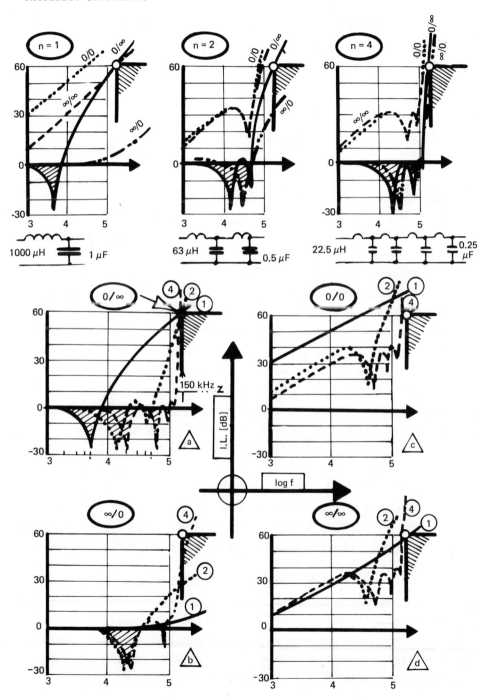

low, but for practical reasons we shall not do too much partitioning anyhow).

The reader will notice that we have not referred at all to the conventional 3 dB cutoff frequency that is used in standard filter theory, as it is completely meaningless for variant mismatch (see Fig. 8.5). Rather, we defined as ($2\pi x$) corner frequency ω_c that frequency for which the filter has a high attenuation a_c, let's say 60 dB, for the boundary condition stipulated.

In Table 8.2 we have listed the eight distinct advantages gainable by partitioning. They can essentially be attributed to the much steeper cutoff and the much reduced value of total L (sum of L's). In Table 8.3, we vary all the critical parameters: division number n and Z_G/Z_L combinations. The results are surprising: We note that for the originally premised $0/\infty$ interface condition, the 60 dB insertion loss at 150 kHz is maintained for all n (n = 1, 2, and 4), as we had stipulated partition. For the simple LC filter (n = 1), the *reverse* interface condition, namely $\infty/0$, causes a quite poor performance. *But this is not the case for higher n.* For higher n, the stopband performance is only insignificantly affected by different mismatch conditions. Conversely, we can also state that the stopband performance is rather independent of filter configurations (pi, T, or L) *and* of mismatch conditions (0/0, ∞/∞, etc.) *if* n > 1 (advantages 1 and 2 of Table 8.2). The reader will also notice the drastic reduction of total L for n > 1 [advantages 4 and 5 (the additional price and size reduction is not generally known), and 6 of Table 8.2]. Partition, then, guarantees good stopband performance, irrespective of filter configuration and mismatch condition. Hence the Z_G and Z_L of partitioned filters are of no great interest at high frequencies, and our data base on interface impedances is needed only for passband and transition band frequencies.

But the diagrams of Table 8.3 also demonstrate the initially perplexing fact that negative I.L. occurs *only for odd* mismatch ($0/\infty$ or $\infty/0$), *not for even* (0/0, ∞/∞) mismatch conditions. Surely, under even interface conditions, there are dips in the insertion loss in the passband, but they do not become negative. Table 8.4

The Solution for Indeterminate Mismatch   213

Table 8.4 Concerning the Theory of Filter Mismatch

THE REAL INTERFACE : See Figs. 8.1 and 8.2 and text

FOR SIMPLICITY SET : or Z_G/Z_o $\Big\}$ $\genfrac{}{}{0pt}{}{\geq 10}{\leq 0.1} \Longrightarrow \genfrac{}{}{0pt}{}{\infty}{0}$
Z_L/Z_o

CONDITIONS FOR INSERTION GAIN (-I.L.) :
Enter sets of extreme values of interface impedances into insertion
loss formula   (see flow graph)   I.L. = 20 log $|x|$:

$|x|$ for		Z_L/Z_o		
		∞	1	0
Z_G/Z_o	0	░░a_{11}░░	$a_{11} + a_{12}/Z_o$	$a_{12}/0$
	1	$a_{11} + a_{21}Z_o$	fully matched	$a_{12}/Z_o + a_{22}$
	∞	$a_{21} \cdot \infty$	$a_{21}Z_o + a_{22}$	░░a_{22}░░

Keeping in mind that each a_{mn} has zeros and that log 0 = $-\infty$,
(negative insertion loss or   insertion gain) we find   :

I.L. negative   for $\genfrac{}{}{0pt}{}{0/\infty}{\infty/0}$ $\genfrac{}{}{0pt}{}{(a_{11}= 0)}{(a_{22}= 0)}$ interfaces only !!!

but not for   0/0, ∞/∞, any/1 , and 1/any interfaces.

TO PREVENT NEGATIVE I.L.:

① Change odd (0/∞ and ∞/0) to even (0/0 and ∞/∞) boundary conditions:
    (a) Lower Z_o, but C may become too large (reactive current!) at 60 Hz or common mode leakage current too excessive.
    (b) Increase Z_o , but larger L may mean too much 60 Hz voltage drop
    (c) Use **lossy L (eddy** currents or partial resistive loading) at low interface; excellent for c.m.
    (d) Add lossy C at high interface if appropriate.

② Shift and dampen eigen resonances:
    (a) Use partition to shift eigen resonances high.
    (b) Dampen eigen resonances (also kills ringing !!!)

provides the explanation for this peculiar phenomenon. If we apply all possible combinations of 0, 1, and ∞ boundary conditions to Equation 8.1, this insertion loss formula becomes rather transparent, as the resulting insertion loss table of Table 8.4 shows. Each of the four extreme interface conditions, *irrespective of filter configuration*, is determined by only *one* of the four a_{mn} terms of the cascade matrix of the filter proper. Now, each a_{mn} (except for the most primitive filter, the single ideal capacitor, or the single ideal series L) has unavoidably one or more zeros, depending on the filter configuration. The numerus of the log in Equation 8.1 reduces to a_{11} for $0/\infty$ and to a_{22} for $\infty/0$ boundary conditions (a_{11} and a_{22} are identical only for symmetric filters). For these *odd* extreme cases, a zero of a_{11} or a_{22}, respectively, means $-\infty$ insertion loss ($\log 0 = -\infty$). In contrast, for the even extreme cases, $0/0$ and ∞/∞, we find the numerus to be $a_{12}/0$, which is not 0 for $a_{12} = 0$ and $a_{21} \times \infty$, which is not 0 for $a_{21} = 0$. Hence there is no insertion gain for $0/0$ and ∞/∞, although there will be dips in the insertion loss. Similar reasoning applies to the in-between cases where one port is terminated by 1 (matched). (Note that a_{11} and a_{22} are dimensionless, whereas a_{12} and a_{21} are impedances or their reciprocals, hence occur always in conjunction with impedances to make the resulting term nondimensional.) We shall now exploit our findings to establish methods that permit avoidance or suppression of insertion gain in an economic way.

8.2.4 Application of Theory to Filter Design

We have now established the following facts.

1. A statistical data base on interface impedances; this means that we have delineated the boundary conditions as they exist-- not less, not more.
2. A transparent theory of mismatched filters for which we can apply these boundary conditions.

It is therefore now rather easy to devise countermeasures against (-)I.L. in the passband (while partition takes care of the stopband behavior). Two principal options are at our disposal (as expressed in concise form at the end of Table 8.4).

1. By various means (listed), we can try to eliminate the causative odd interface conditions by converting them to even conditions. The preferred choice is (1c): Change 0 into ∞ by adding a lossy L and consider it as part of the Z_G or Z_L, whichever is 0. This approach is particularly beneficial for common mode filters, for which bifilarly wound coils on ungapped cores (high μ, no bias effect) are of small size. The normal mode impedance of such coils is 0, hence causes no voltage problem at 60 Hz (except IR drop). Since very large L's can be achieved by such ungapped cores, common mode filters can be made with small total capacitances, assuring minimal 60 Hz common mode leakage current (often required to be below 0.5 mA).

2. The second option (Table 8.4), the damping of eigenresonances (if methods available under item 1 are not applicable and if we are concerned about ringing) is more complex and requires amplification. We quickly skip over Figure 8.6, which permits the quick calculation of inductors made lossy by heavy laminations. Rather, we address the problem of using resistors (preferably carbon composition) for damping resonances.

Example 8.3 Checking the Theory. Here we are concerned with a combination of applied theory and quasiempirical work (for proof).

The purpose of Figure 8.7 is to substantiate the importance of the (characteristic) impedance level as it affects interfacial *and* eigenresonances. Four LC filters, each partitioned with n = 3, are juxtaposed. All have 60 dB insertion loss at 150 kHz in a 50 ohm/ 50 ohm system. The differences are as follows.

Filter A: Total C = 40.0 μF,
           Total L = 1.0 μH; Z_o = 0.16 ohm
Filter B: Total C = 10.0 μF,
           Total L = 5.8 μH; Z_o = 0.76 ohm
Filter C: Total C = 2.0 μF,
           Total L = 50.0 μH; Z_o = 5.00 ohm

216   8. Filtering for EMC

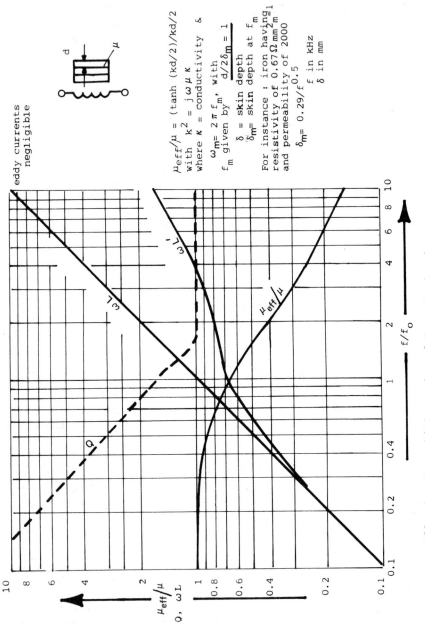

Figure 8.6 Effective Permeability and Q of Heavy Laminations

The Solution for Indeterminate Mismatch

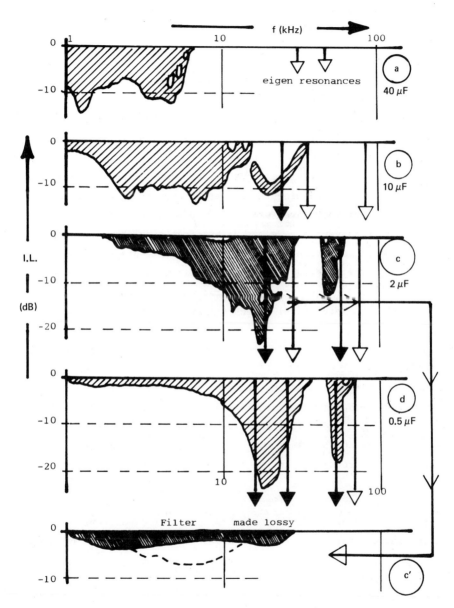

Figure 8.7 Effect of Z_o and Q on Insertion Gain

Filter D: Total C = 0.5 µF,
            Total L = 275.0 µH; Z_o = 23.00 ohm

These four filters determine four sets of a_{mn}, which we enter into Equation 8.1 (Fig. 8.4, together with all normal mode boundary conditions [Fig. 8.1(a)-(c)]). From the computer readout, we sketch only the insertion gain, since it is the only concern for the moment. We also plot the eigenresonances (vertical arrows), which can occur for all possible extreme cases.

What do we find? Filter A, the very low Z_o filter, shows no negative I.L. for the eigenresonances, as we expect according to our theory, but there are, well below 10 kHz, quite a number of interfacial resonances. We may accept them since they involve only rather modest insertion gain. Now, the more we proceed from A to D, the more the interfacial resonances disappear and the eigenresonances become stronger. They are well above the intensity we observed for the interfacial resonances at A, where Z_G resonates essentially with the 40 µF in the first approximation (see Table 8.5). Again, our theory is proven, but we still have to get rid of the eigenresonances. Now it is much easier to dampen higher resonances than lower resonances; thus we select filter C. For damping we select the scheme depicted in Figure 8.8, which gives us an indication of the order of R and C required. For inductive damping see Figs. 8.6 and 8.8 and Problem 97, Chapter 12. Computer work (Eq. 1), with corresponding matrix terms inserted, quickly shows, by trial, that, for example, a (C_1/C_2) = 0.3 and R = 16 ohms gives the quite acceptable results shown in Figure 8.7 as (c'). Thus we have to add to each of the feed-through filter capacitors of 2/3 µF a 16 ohm resistor in series with 2 µF. If we use very short leads, the additional C_2 does not have to be a feed-through configuration (price), since C_1 is dominant only for higher frequencies.

Figure 8.9 contrasts the contours of the insertion loss of filters (c) and (c') (lossless and lossy). This ends Example 8.3, which brought us a clear understanding of the mechanism that causes resonances in and near the passband.

Table 8.5 Interfacial Resonances

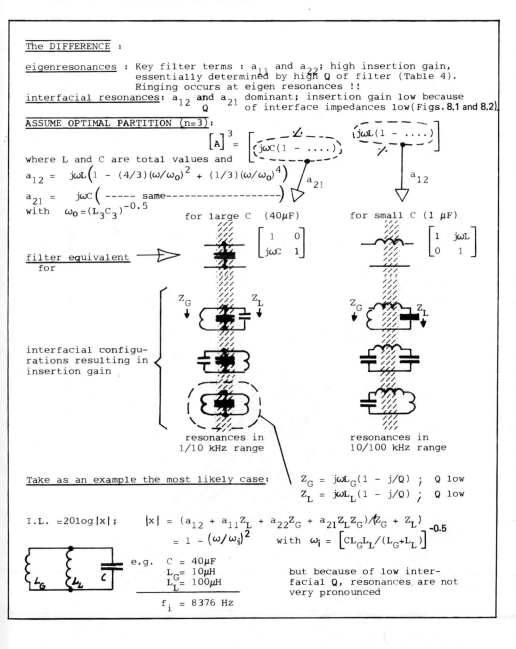

8. Filtering for EMC

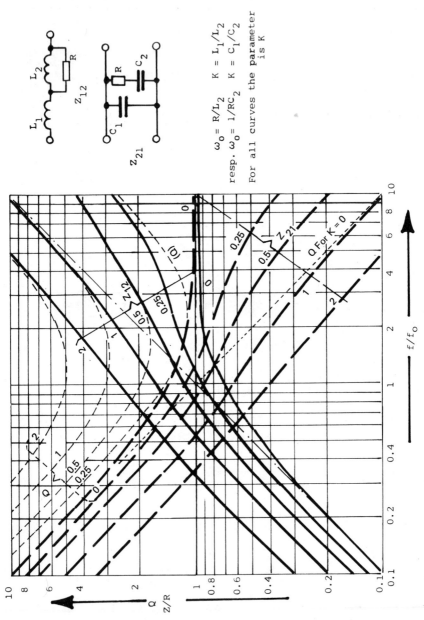

Figure 8.8 Resistive Damping of L and C, Normalized

The Solution for Indeterminate Mismatch 221

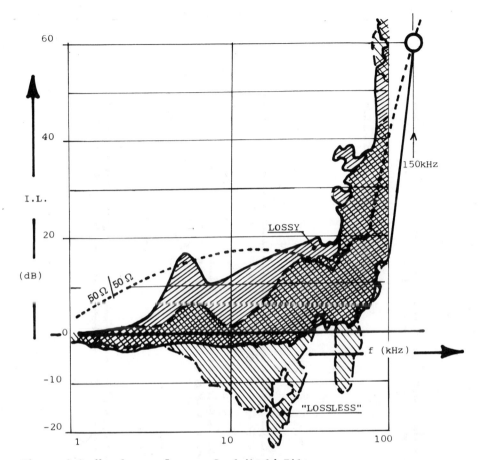

Figure 8.9 How Losses Improve Real-World Filters

For convenience of reactance calculations, we add the universal LCFX nomograph (Fig. 8.10), which consists of two interleaved systems: the coarse dot system to determine the exponent of 10 multipliers and the fine log system to determine the digits themselves.

Example 8.4 Use the LCFX nomograph to determine the reactance and required L to resonate 5×10^3 pF at 1.5×10^4 Hz. Answer: On the coarse scale with C = 3 and f = 4, we read X = 3, L = 4 and on the fine scale with C = 5 and f = 1.5, we read L = 2.2 and X = 2.1. The answer, then, is: L = 2.2×10^4 μH, X = 2.1×10^3 ohms.

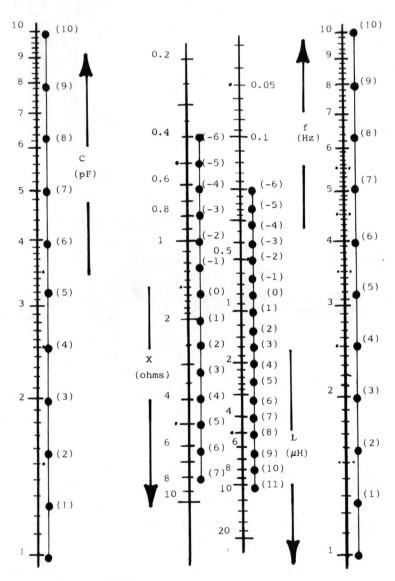

Figure 8.10 LCFX Nomograph

The Solution for Indeterminate Mismatch 223

Let us summarize the potential dysfunctions of filters and their causes and remedies:

1. *Component resonances*, creating "suck-out" points high in the stopband and caused by the nonideality of lumped and distributed-parameter components (Table 1.9), can be dampened or shifted to unininterestingly high frequencies by:
   A. Partitioning (smaller L and C)
   B. Deresonated ceramic feed-through capacitors
   C. Ferrite beads or shells
   D. Lossy pieces of line (mix of powdered ferrite or iron, carbon, and plastic around a coiled conductor
   E. What else?
2. *Eigenresonances* are caused by mismatch in the very frequency range, and thereabouts, where under matched conditions the 3 dB point lies. They cause
   A. Negative insertion loss = insertion gain that may amount to 35 dB at odd interface conditions only.
   B. Ringing, when Z_G or Z_L is switched (see Fig. 1.12). The remedy: damping of these resonances.
3. *Interfacial resonances* are resonances between the filter and its non-0 or non-∞ Z_G and/or Z_L. They lie below the eigenresonances and are usually negligible because of the low Q's of the interface impedances.
4. *Poor stopband performance* caused by the wrong, nonpartitioned filter configuration, is remedied by partitioning.
5. *Conventional lossy filters* are distributed-parameter, high-frequency-cutoff LP filters made lossy in the stopband. They prevent component resonances. But they do not work at low cutoff frequencies, where their loss mechanisms do not work [2].
6. *Meaningful testing* for conditions listed above (see Chapter 11).

We add two more (self-explanatory) diagrams helpful for filter design: Figure 8.11, pertaining to common mode configuration (FATT-MESS), and Figure 8.12, pertaining to the limitations of ferro- and

8. Filtering for EMC

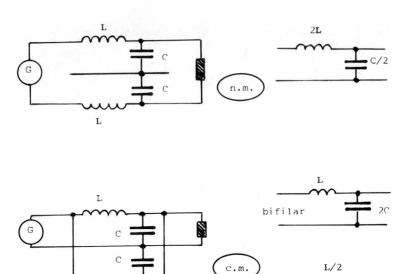

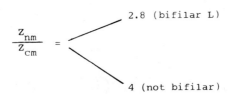

Figure 8.11 Normal Mode and Common Mode Filters

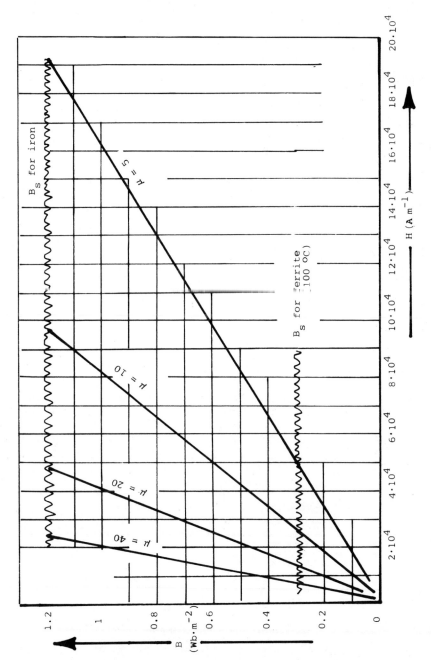

Figure 8.12 Permissible Permeability as a Function of Current.

ferrimagnetic materials under current bias (FATTMESS). It permits us to read the maximally permissible effective permeability as a function of the biasing current [see Table 1.4(a2)].

REFERENCES

1. Assuredly Effective Filters, H. M. Schlicke, IEEE Transactions on EMC, August, 1976, pp. 106-110.
2. Based on pioneering work by F. Mayer, LEAD, Paris, France, such filters exploit the HF losses of ferrites and are widely used as automobile ignition cables for suppression of VHF and UHF radiation.

9 Grounding and Wiring, Continued
You Must Plan Ahead

Throughout the course, but particularly in Chapters 1 and 5, we have met obstacles, constraints, and uncertainties in conjunction with grounding and wiring, which are supposedly very simple things. But they lose all simplicity when the many-purpose, many-function conductors and fields intermesh, since then they can no longer be considered independently. To make this highly interactive situation manageable, we introduced systemic partition and isolation.

Such preventive planning stressed the main principles involved (Chapter 5). In our effort to bring out the essence without much distraction, we left out some important details, which we discuss now.

9.1 EQUIVALENT LINES

Table 9.1, top, juxtaposes three equivalent line configurations having identical L', C', and Z_o. These relations are helpful in normal mode considerations. Table 9.1, bottom, permits calculation of the effective (equivalent) radius of wire bundles operating in the common mode (see the example).

Table 9.1 Equivalent Lines

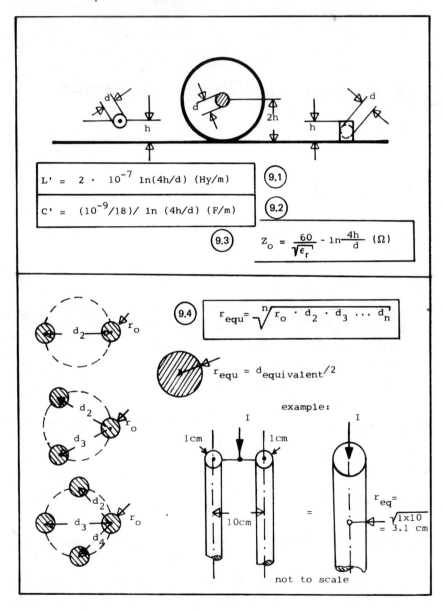

9.2 MORE ABOUT VOLTAGE GRADIENTS

Voltage gradients (caused by lightning or ground faults) on the surface of the ground are of great concern because of the dangerous step voltages they can create. We discussed simple geometries in Sections 1.2 and 1.3. Here we want to demonstrate (1) how step voltages can be reduced by proper configuration and (2) how they may inadvertently be enhanced by disregarding neighboring conductors.

In Table 9.2 we develop--from (a) to (d), from the worst case to the best case--how to make the step voltage negligible. We will use numbers to get a better feeling for it.

In Table 9.2(a) the wire just "touches" the ground: momentarily we get enormous gradients, resulting in fireworks. In (b) we shove the wire 10 m into the ground and still get intolerable field strength, particularly for ρ = 1000 ohm-m. In (c) we are quite a bit better off. Here we apply a circle of 10 m radius. Exactly, the calculation refers to a 10-m-radius sphere which we approximate roughly by building a 10-m-radius circle of small grounding rods. But the best solution is (d), where we apply guarding in conjunction with a larger radius. The resulting voltage difference between points A and B is insignificant.

In Figure 9.1 we contrast (a) field strength enhancement and (b) field strength reduction. Concerning the latter: We can reduce the step voltage the deeper we insulate the ground wire under ground (as indicated).

Figure 9.1(a) is in principle applicable to all cases where a grounded, but not bonded, object (a lake, river, gas tank, etc.) is located rather close to a ground rod into which a current is being discharged [1]. At point A, the closest point of the lake, the step voltage is much greater than it would be if the lake were absent.

Table 9.2 Comparison of Grounding Arrangements

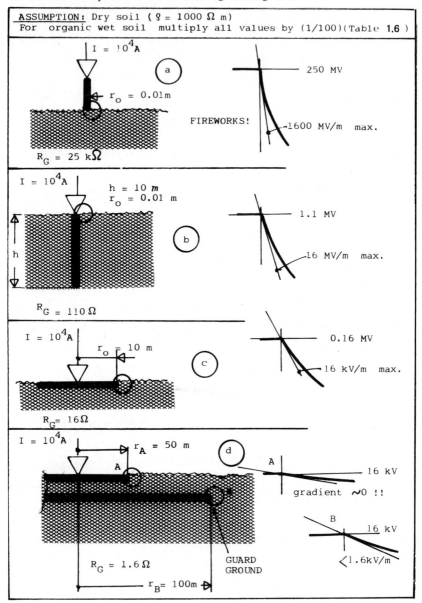

More About Voltage Gradients

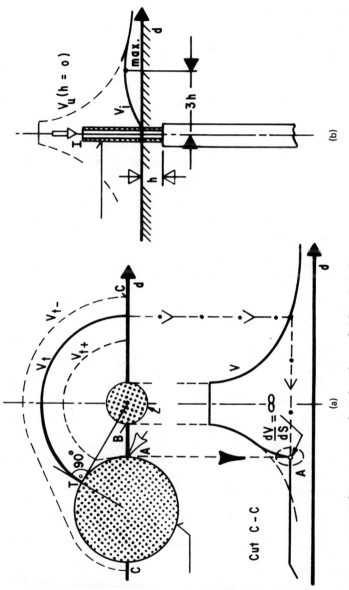

Figure 9.1 Enhancement and Reduction of "Step" Voltage

9.3 KEY POINTERS FOR ARCHITECTS AND CONTRACTORS

To make a facility safe from hazards and EMI, it must be carefully planned in many details from the very beginning. Adding shielding and grounding or changing wiring after the building has been completed can be a very costly affair. Also, maintenance can be very difficult and expensive unless projected from the start.

In keeping with a major objective of the course (i.e., to stress principles and to resolve dilemmas), we have no intention of going into all the procedural specifities required by codes and "good engineering practice." The reader will find Ref. [2] a good and well-organized source for the many practical considerations required in facility planning. Admittedly, I do not agree with the authors about the effectiveness of "control common" (there called signal grounds), but instead, suggest the rigorous use of partition and isolation (Chapter 5). Nevertheless, Ref. [2] is a very worthwhile document to have, as it goes into many practical details, such as corrosion prevention.

Here we want only to compile the key provisions that the architect and contractor must plan into the facility to avoid costly retrofitting.

1. *Site selection.* Check for areas of heavy industrial zoning and plans for radio and TV stations, electrical railways, high-tension lines, and so on, as measurements made at the projected site today may be obsolete tomorrow (without legal recourse for you). Thus the following comments contain points that permit easy complementation of your facility into a rather reasonable shielded structure without holes--electromagnetic holes, that is.

2. *Structure and metallic elements (bonding)*
    A. Include the metallic structure of the building in lightning protection scheme.
    B. Connect-through all long metallic parts (e.g., raceways).
    C. Make sure that all rebars are securely welded to form continuous cage wiring.

D. Floors and ceilings should be made part of continuous cages by cross-connection via mesh wire or rebars to vertical metallic elements such that each floor is a wired cage.
   E. In particular, provide continuity of the metallic network of the structure with metallic window frames (and doors) such that shielding of windows can be conveniently added if necessary.
   F. Provide a shielded room for very sensitive equipment.
   G. Make sure to bond all larger metallic objects to the grounding system using the star system (tree).
   H. Water pipes, gas pipes, and so on, should be part of the ground system, *but* disconnect them from neighboring systems by dielectric unions. Provide corrosion protection and avoid field concentrations.
3. *Electric wiring*
   A. Follow codes and apply protective devices.
   B. Use fiber optics to eliminate EMI and hazards and drastically reduce installation costs. At least use electrooptical isolators in signal and control lines.
   C. Use isolation transformers.
   D. Put fluorescent lamps and other noise-generating devices (hand tools) on another phase/neutral than very noise-susceptible devices.
   E. Filter power feed lines.
   F. Always expect a voltage difference between two different ground points.
   G. Use ground fault interrupters (GFIs) where dangerous leakage can occur, in particular in wet surroundings.
   H. Reread Sections 5.2, 5.3, and 5.4.

REFERENCES

1. Erhoehung der Schrittspannung an See-ufern, L. Hannakam and T. Hermes, ETZ-A, 1973, pp. 505-508 (in German).

2. Grounding, Bonding, and Shielding Practices and Procedures for Electronic Equipments and Facilities, Vol. 1: Fundamental Considerations; Vol. 2: Procedures for Facilities and Equipment, H. W. Denny et al., 1975, Final Report, Report No. FAA-RD-75-215 (available from National Technical Information Service, Springfield, Va. 22161).

10 Standards and Truths
Use the Code and Use Your Head

See also Sections 1.3, 5.2, 5.3, 8.1, 11.1, 12.2, and 12.4.

Throughout the course we encountered difficulties with standards. Why is this so?

10.1 BASIC CLASSES OF STANDARDS

There are basically two kinds of standards:

1. *Normative standards (deterministic conditions).* (Typical ones are measuring units and specific design norms.) Their objective is to prevent chaos and to facilitate exchangeability of hardware and information. Normative standards are "absolute," undebatable, unconditional, unambiguous. Nobody quarrels about them because they have nothing to do with statistics, rather are clearly deterministic.

2. *Limitative standards (probabilistic conditions).* (Typical are safety codes.) They are principally different in that they have to cope with random distribution. We have essentially two choices:

    A. Using the *statistical means* or something close to it. This is the most primitive approach. Here we cannot think like life insurance underwriters, for whom averaging large numbers is fully adequate. Rather, we must think like a person buying life insurance. The thing we are concerned about is an individual, not an average. (See Section 8.1 for an excellent demonstration.) Also, a normalized standard pulse falls in this category. It is not a normative standard as defined before. When we stipulate a typical or statistical mean of the myriad of pulses encountered in a system, or worse in many systems, we create an artifact that may or may not be meaningful under probabilistic conditions. But it is certainly not very sensible for determining criticalness.

B. Delineating *realistic worst-case* conditions. This is the most reasonable, but also the most difficult method of standardization. It is premised on a good, statistical data base and a good theory that grasps the essence of sets of situations. For economic reasons, we must make sure to make the standard *not* for *any imaginable* worse-case condition but for *realistic* worst-case conditions. See the replacement of the 50/50 ohm method by the 0.1/100 ohm, and the opposite, the method for determining the insertion loss of power feed line filters working into unavoidably indeterminate interfaces (Chapter 11). Or in the case of safety standards, we can use statistics to arrive at safe threshold limits *if* we have a sufficiently broad data base.

10.2 LIMITS INHERENT IN STANDARDS

1. Standards are not sacrosanct. At best they present the best present knowledge. A good limiting standard is periodically updated to include new findings that may be so fundamental as to require complete rewriting of the standard.
2. In rapidly evolving technologies, it makes no sense to issue standards lest one wants to stifle the industry. Tentative guidelines are the best one can do as long as everything is still very much in flux.

3. As discussed in Chapter 5, standards essentially address specific requirements to permit unambiguous action by average people "skilled in the art." But in complete systems, we encounter nonlinearities and multiparameter interactions, often invalidating accepted statistics. Thus, as we did in Chapter 5, the clash of standards can be undone only by partition and isolation.
4. To make them usable for the average practitioner, standards are sometimes so oversimplified that they should be discarded.
5. In so-called "voluntary" standards, it has happened that limits were set to exclude competition.
6. "Recipe" engineers and administrators "going strictly by the book" are unable to accept limitations of standards. Such people should be considered as what they really are: people of very limited abilities.

10.3 THE THREE TRUTHS

In the judicial realm, we have many laws and recorded precedents, but we still need judges and lawyers because the cases require legal thinking to establish relevancy and truth.

Similarly, in EMC engineering, we have many standards, but we still need systemic thinking. The similarity exists in that in both fields, considerations must be made in the context of the whole situation. But there the similarity ends. The engineer should become very aware of the striking difference between legal thinking and engineering thinking. If called as an expert witness by a party, the engineer is expected to think differently.

Legal people have two kinds of "truth":

1. The judge, supposedly full of insight, looks for the *technical truth* in terms of *absolute findings* based on facts, in accordance with specifications and standards, provided that they apply.
2. The lawyers of either party try to establish the *adversary truth*, each his or her own. The proponent for a proposition selects only these facts that aid his or her case. And the

proponent against the proposition selects only detrimental facts and argues affirmative facts to be irrelevant. The judge, looking at both sides, decides the truth.

In contrast, the engineer is looking for the *engineering truth*, stating his or her finding of facts and including all uncertainties, conditionalities, assumptions, approximations, and counterindications.

The reader should refer to Example 1.13, where the key question is: Is a specific standard applicable or not? Or put another way: An imperfect human-made law (standard) cannot compete with the basic laws of nature. Standards are never a substitute for thinking.

10.4 FRAGMENTATION OF STANDARDS

Dissatisfaction with and disregard for insipid standards are justified. But dissatisfaction and impatience with the fragmentation and imperfection of the vast majority of standards is often not justified. Standards must be understood being part of life which is fluid and not predeterminable. If the need is really felt strongly, the standard will be developed or modified out of self-interest of the people concerned. But naturally, the larger the mass of the people, institutions and/or countries involved, the slower the reaction and often the more watered-down and unfirm is the standard. We must not discard the need for an ideal in view of the imperfection of its realization.

Concerning specific standards, see Chapters 5, 6, 11, and 12. For an excellent survey of international and national standards, see Ref. [1] which lists the CISPR standards and addresses of all the important national and international standard agencies. VDE Interference Regulations of West Germany are discussed in Ref. [2].

REFERENCES

1. National and International Radio Interference Regulations for Consumer Products, H. K. Mertel, Record of the 1976 International IEEE EMC Symposium, pp. 23-28.
2. VDE Interference Regulations of West Germany, H. K. Mertel, EMACO, P. O. Box 22066, San Diego, Calif. 92122.

11 Comments on Measurements
Most-Needed Corrections Only

For our amendatory purposes, we divide EMC measurements into three groups:

1. Test methods developed for military/communications systems. These are numerous and mostly good to very good, but often not applicable to our civilian systems, as they are concerned primarily with narrow-band emission or susceptibility.
2. The much less numerous, and often still embryonic test methods specifically developed for the systems we consider here. Professional and trade organizations now put more emphasis on developing methods and specifications concerning impulsive noise (Chapter 5). Here we shall discuss only impulse testing of shielded subsystems.
3. Test methods, taken from group 1, that would and should be applicable *if* they would make sense. Luckily, there are only a few of miscreations. We shall correct the worst of them: worst-case filter measurements.

It is not within the scope of this book to discuss established test methods. Most equipment manuals (particularly on spectrum analyzers and storage scopes) are written very well. Rather, we restrict ourselves here to the most salient cases cited under groups 2 and 3, presenting the most pressing problems.

11.1 AT THE EDGE OF THE STATE OF THE ART

11.1.1 Impulse Testing of (Sub) Systems

Reference [1] describes briefly the relay noise and the motor noise tests. They are brute force methods used to check the shielding and filtering effectiveness of sensitive systems or subsystems. The tests are made at elevated temperatures (70°C), where the switching speed of the chips is higher. The full adequacy of such tests is doubtful, as very fast transients or rise times are not generated. Their high-frequency components would easily pass through imperfections of the shields or filters of the circuits under test.

It seems, therefore, advisable to subject the shielded system to fast-rising pulses. To do so for large systems would require rather high energy. Figure 11.1 depicts the schematic of an ingenious pulse generator particularly suited for our purposes [2]. Such pulse generators are smaller and less expensive than Marx generators. The unique features are: A water capacitor (ε = 80) in the form of a rather small coaxial line; a copper sulfate load that can be easily varied to change the stepped fall of the pulse; high-pressure switching spark gaps; 5 s bursts of pulses with 5 ns rise time can be created. Such pulse generators were originally developed for EMP testing in Europe, where more modest financial resources are a dominant constraint. It is my understanding that such cost-effective test generators can also be built in the United States.

We now go back to our primary objective: With such steep-rise-time pulse generators,

We can feed broadband (Chapter 2) antennas to radiate into the system under test.

We can feed wide-cross-section transmission lines (Crawford cells) into the fields in which we place our test object with the polarities desired.

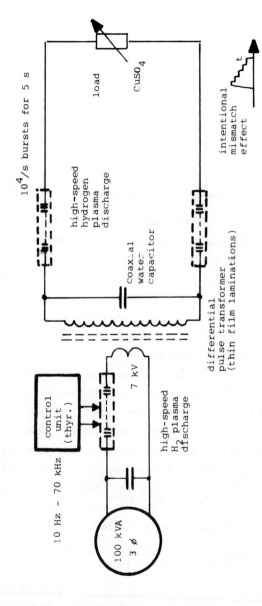

Figure 11.1 A New, Economical Pulse Generator

We can also modify them such that the equipment under test forms part of the outer conductor of the line [3].

In either case, we do not have to be concerned with standing waves in the shielded enclosure needed about the whole test arrangement; thus there is no need for microwave absorbers or mode-stirred chambers (see Sec. 7.1.1.4).

Admittedly, this discussion was quite sketchy, as things are still in development. But in view of the excellent works (Refs. [5] and [6] in Chapter 1) on established technology, these hints at the edge of the state of the art should suffice.

11.1.2 Rational Worse-Case Filter Measurements

Now, at the end of this course, we come to the solution of the singularly notorious EMC predicament that is long due for a fundamental correction. Chapter 8 has made clear that the conventional measurement of power feed line filters in a 50/50 ohm interface system (Mil. Std. 220A)--*if* the filters operate, as they very often do, into indeterminate and variant interfaces--is meaningless, in fact nonsensical. One does not measure something under completely uncritical conditions contrary to fact.

But, still, some filter manufacturers adamantly defend Mil. Std. 220A and other inadequate methods (see Sec. 8.2.1). The conventional methods used for years have conditioned many EMC engineers into accepting them as reasonable because it has been done for so long. To maintain their comfortable status quo, some (only) filter manufacturers go to quite some length in their polemic against change. But the facts are catching on. And needed changes are, by far, not as "bad" as imagined. In this context we have to mention that some alert filter companies are already using low-Q inductors to reduce ringing and insertion gain.

The IEC (International Electrical Commission), in clear recognition of the inadequacy of Mil. Std. 220A and similar nonrealistic methods, is finalizing (according to R. Showers) realistic test methods for power feed line filters. Possibly with some minor modifications, the IEC document will present two alternative methods:

Method A. Based on a British Position Paper (condensed in Ref. [4]). The Thévenin and the transfer impedances are determined.

Method B. Based on a U.S. Position Paper (a condensation of Chapter 8). The filter is measured in a 0.1/100 ohm and 100/0.1 ohm interface system, from 2 kHz to at least 200 kHz for a 150 kHz filter. One realization, particularly suitable for measuring under current bias, is depicted in Figure 11.2(a). Figure 11.2(b) shows a very simple version of testing for ringing only.

The commonalities and differences of approaches A and B, including their assumptions and inferences, are summarized in Table 11.1.

These alternative methods are meant as *acceptance tests*. They are simple and predictive, and most of all they solve, on a sound basis, a long-standing EMC dilemma about which, thus far, there has been much discussion, but no answer or action.

As already stated, some filter manufacturers are against--very much against--the new methods. The reasons are understandable, although not justified: Their filters will not meet the new standards because they are arbitrarily optimized for 50/50 ohm interfaces.

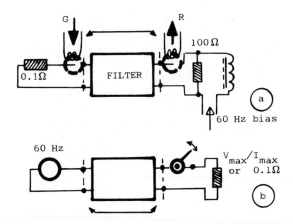

Figure 11.2 Testing of Filters: (a) for Worst-Case Conditions; (b) for Ringing Only

Table 11.1 Juxtaposition of the U.S. and British Approaches and Their Findings

No.	Question	Approach B	Approach A
1	Mil. Std. 220A	Meaningless for low and medium frequencies.	Meaningless.
2	Assumptions	No assumptions; rather, broad statistical data base on "all" Z_i's encounterable in 60 Hz systems (including Q's; normal mode, common mode). Confined to real world.	$Z_G = 0$, "any" Z_L; total impedance plane. British aircraft systems.
3	What measured and how	Insertion loss directly; measuring filter in 0.1/100 ohm (and reverse!) system from 2 to 200 kHz.	Attenuation, indirectly: measuring transfer impedance and Thévenin impedance.
4	Filter types	Pi, T, L, reverse L, and multiples.	Multiple pi and T.
5	Critical passband behavior	Eigenresonances for $0/\infty$ and $\infty/0$ conditions resulting in up to 35 dB insertion gain are considered critical, interfacial resonances of secondary significance because of damping exerted by low Q's of Z_i's (data base).	Resonance between filter and load considered critical (apparently high Q's of load assumed): negative attenuation predicted.
6	Recommendations for good passband performance	Dampen passband resonances by introducing losses in filter.	Dampen passband resonances by introducing losses into filter.
7	Good stopband behavior	Multisection filters.	Multisection filters (by implication).

At the Edge of the State of the Art   245

Yet filter manufacturers will greatly benefit if their filters meet upcoming standards because their products will then work predictably and will probably be smaller and cheaper. The reasoning of such filter manufacturers is illustrated by a typical question: "Why should I design a filter for the 1% of bad cases if it costs much more?" Well, the facts are:

1. Indeterminate conditions, caused by switching and alterations, particularly in many-filter installations, are much higher than 1%. The worst case is often the normal case.
2. The specification writer no longer has to add 40 dB or so as a "safety margin," as may be done fallaciously if one does not know whether or not the filter will work (brute force approach = waste).
3. See Table 8.2 concerning cost reduction achievable by partitioning.
4. For a single-filter, defined-interface condition, the worst-case approach is certainly superfluous; rather, the best suited filter configuration should be applied (see Chapter 8). If, in such a situation, switching occurs, a lossy filter (lossy in the pass- and transition band!) is required.
5. Mil. Std. 220A is completely fitting for
    A. Quality control testing, *provided* that representative samples of the filter have been successfully acceptance tested for worst-case conditions.
    B. *If* the worst-case condition has been met for the passband and transition bands (say 1 to 200 kHz for an a_c at 150 kHz), measurement of the filter under 50/50 ohm conditions above 200 kHz is fully acceptable even for acceptance testing. (Why?) This statement is not made to placate nervous filter manufacturers but is the corollary of filters designed for worst-case conditions as defined by method B.

REFERENCES

1. Getting Noise Immunity in Industrial Controls, H. M. Schlicke and O. Struger, IEEE Spectrum, June, 1973, pp. 30-35.
2. Private communication with Dr. F. Fruengel, President, Impulsphysik GmbH, 2 Hamburg 50 Risen, P.O. Box 56 0160, Germany.
3. The Transmission Line Technique for the Measurement of Radiated Susceptibility, R. E. Hartman and W. A. Kesselman, Record of 1978 IEEE International Symposium on EMC.
4. Worst-Case Suppressor Methods--The Minimum Attenuation Concept, M. L. Jarvis and J. M. Thomson, IEEE Transactions on EMC, May, 1977, pp. 99-100.
5. Comments on Ref. [4], H. M. Schlicke, IEEE Transactions on EMC, May, 1977, pp. 100-101.

12 Problems and Solutions
Learning to Co-think EMC

All through the foregoing chapters we have employed a large number of numerical examples to give the reader a feeling for the order of magnitude involved. Important as such amplification is, it is still nothing but well-prepared exercises (problems defined and predigested solutions). This helps to conceptualize engineering laws under highly idealized assumptions very seldom found in actual systems. But knowing all engineering laws, even in your sleep, does not necessarily include engineering thinking, which, in addition, requires keen perception, judgment, and mental flexibility. Engineering tries for cost-effectively meeting many contrary objectives and constraints--hence our emphasis on the systemic approach and the multifacedness of coexisting criteria as expressed in the FATT-MESS acronym. Another circumstance poses a further mental challenge: the dominant fact that the primary mission of the planning is the system under consideration, for which EMI is only a feared, often considered to be debilitating plague, of somehow foggy origin and behavior, of even more foggy tractability, attempted by empirical fogies. Hence we stressed the need for EMC co-planning by the systems designer instead of naively applying isolated EMC standards by so-called EMC experts after the system has been built.

To do a good engineering job, the systems designer must not only have a critical eye for questionable standards but must also be good in thinking beyond engineering. Strict engineering thinking is often the greatest drawback in an engineer's thinking. An engineer must realistically include human and social implications and ramifications in all engineering considerations. The following two sections treat very briefly two aspects of such real engineering which is unfortunately not taught in schools.

12.1 ON DEFINING A RAW PROBLEM

When faced with any raw set of interactive problems containing dilemmas of constraints, uncertainties, and obstacles, the analysis of the problem predisposes its solution or lack of solution. Three generic ways of looking at a problem to gain perspective are at our disposal. We must involve all three basic ways to get to the crux of a complex matter. Here are the three ways: One can analyze a "mess" by making (A) three basically different assumptions based on (B) three different mental attitudes constituting (C) three different problem approaches, which, if taken alone, cause (D) three different insufficiencies or dangers.

The three fundamental approaches are:

1. *BACKWARD-LOOKING*
    A. Everything is orderly and linear (cookbook engineer, accountant, lawyer, administrator).
    B. How do I establish or reestablish routine?
    C. Asking "What is the applicable code or convention?"
    D. Oversimplification, missing real issue or dangers.
2. *FORWARD-LOOKING*
    A. Everything is in flux and nonlinear [(engineering) leader, innovator, entrepreneur].
    B. What are the opportunities and challenges?
    C. Asking "How does the whole thing work as a system?
    D. Instability, threat to established order.

On Defining a Raw Problem   249

3. INSIDE-LOOKING
    A. Everything has basic human springs (leader, self-actualizer,
        but also politician, salesperson).
    B. What is the basic need? What makes the other person tick?
    C. What are the facts in terms of the other person's (or my)
        values?
    D. Purely political decision disregarding bad effects.

It is obvious that looking at a "mess" from only one of these
generic angles is insufficient. Rather, all three basic aspects
must be considered concurrently. Intuition and judgment, logic and
calculation must be combined. Naturally, we can only hint here at
the initial approach to such comprehensive thinking. We cannot go
into any detail here on how to achieve such mental effectiveness,
which is needed for the upper percentiles of any profession.

Particularly concerning EMC, the engineer should look at the
raw input from all possible angles, to extract the real issues so
as to make things tractable. Here the engineer must have a good
data base of pertinent facts and relations as a necessary, but not
sufficient prerequisite. And a good data base, which we tried to
establish throughout the book, means organizing what is useful
information and identifying and throwing out blatant misconceptions
and fallacious oversimplifications.

But the best data base is of no avail unless we keep in mind:

1. A definition is given by (a) the *genus* (the next higher order
    of a classificatory system) and (b) the *differentia*, which
    distinguish it from other species of the same genus. AND: the
    definer has always the choice of genera. He or she should
    select not only one; in fact, the definer should even include
    the opposite or the "not-so."
2. In an interactive system, a change or disturbance at point A
    may cause another, very unsuspected disturbance at point B
    (see, e.g., Sec. 1.3). We can expect to invalidate such

counterpositive effects if and only if we EMC-order a priori the EMI disorder inherent in any non-EMC-co-planned system. That means, according to Chapter 5, organizing by partition, isolation (and redundancy), with a heavy emphasis on differentiation.
3. Any of the eight FATTMESS criteria can throw you off course. So check for contingencies for each of them.
4. In dealing with EMI, as in all controversial matters, we have to deal with people and their constructs. And we must be aware that there are all kinds of people. We must forget what they say and determine what they really mean. We next make some pertinent comments on that.

12.2 COPING WITH ALL-TOO-HUMAN FRAILTIES

We saw in Chapter 1 (the example without number and the sailboat accident) that "human" aspects enter significantly into many an EMC consideration (though no textbook will admit that). In Chapters 5 and 11 we demonstrated the all-too-human oversimplifications and pitfalls of standards. We also warned that the selling of EMC devices or services is sometimes, but by far not always, done for the real benefit to the seller at the expense of only an imaginary benefit to the buyer. Thus we have compelling reasons to acquire a good foundation in EMC principles.

To illustrate, let me relate two simple instances which I can easily document. They refer to aspect 3 (delineated in Sec. 12.1) in conjunction with filter standards, for which the engineering facts and inferences are discussed in Chapters 8 and 11. The first instance: When I submitted the manuscript "Assuredly Effective Filters" (the basis of Chapter 8) to the IEEE Transactions on EMC, it was flatly rejected by the reviewers. Fortunately, the editor, Dick Schulz, recognizing the significance of the paper, had the guts to print the paper. Although he could not give me the names of the reviewers, he admitted upon questioning that the manuscript had been submitted to engineers from filter companies for review, in the assumption that they were most suitable for the purpose on hand. Need I say more?

The second instance: I quote verbatim from page 33 of Ref. [1]: "Garlington (Sprague Electric Co.) noted that filters are designed to work into specific impedances, but they are frequently used with a variety of source and load impedances (which may not be known). This can result in a filter not providing the required isolation between the source of conducted interference and the victim system." End of quote, constituting, in my opinion, a statement of poor engineering or, at least, of a peculiar disregard of customer needs.

In Chapters 8 and 11, we gave the realistic, economic, and scientifically sound solution to this "hot potato" dodged by government bureaucrats like a dangerous, contagious disease. Some U.S. filter manufacturers are still fighting, in ways I would not use, prospective international and national standards pertaining to worst-case filter measurements.

But although then left without useful official standards, the filter users have at their disposal two simple tests to ensure that a filter will operate well even if its interfaces are indeterminate and variant. They are: (1) Is the filter a multisection filter? A yes answer is required. (2) Does the filter strongly ring when I switch a short circuit on and off quickly? Here you need a definite no answer. Although "conservative" filter manufacturers still peddle their outdated wares and so doing continue to give conventional EMC a black eye, there are a number of progressive filter makers who introduce losses in the transition band to reduce drastically ringing and insertion gain.

Other standards may be no good because of careless omission of critical details. For instance, see Problem 8 pertaining to far-field measurements according to VDE standards. A hidden ground plane is not mentioned, so that tests are not reproducible unless a ground plane (metal grid) is provided.

12.3 PROBLEMS, PROBLEMS: IMPROVE AND CHECK YOUR UNDERSTANDING

As one learns best by doing (which includes thinking), a desirable hands-on experience could be accomplished the fastest by a carefully planned laboratory training course. Such a course is now being studied in conjunction with a carefully selected, progressive EMC company; but its realization is still some time off. In the meantime, we do the next best thing available: provide a large variety of problems encountered in practice. The mixture of problems is put into three groups:

Group A. Simple multiple-choice questions. Do not get conceited if you answer correctly 50% of the yes/no questions. Anyone can do that, given enough questions of this kind.

Group B. A simple answer is sufficient, be it a sentence, sketch, or simple calculation.

Group C. A bit more complex problems. Essentially digitally trained engineers, even the best ones, may find these mostly analog problems disturbing. But, after all, the equipment interferes in and with the real world, which is an analog world from where come most of the disturbances we want to invalidate.

Many problems stem from questions asked during Continued Education EMC courses sponsored by the IEEE. They are purposely not ordered according to the chapter to which they belong. Their random distribution tries to approximate the actual EMI world, where the problems are also not prearranged and predigested.

12.3.1 Problems: Group A

1. The relation $E/H = 377$ ohm is correct only for the: near field; far field.
2. It makes no difference where a good RF shield is grounded if no low frequencies are involved. True; False.
3. Can a fast, large system be designed such that there are no ground potential differences between locations? Yes; No.

Problems: Group A

4. Is a guard shield good at high frequencies? Yes; No.
5. Because a ferroresonant transformer is based on nonlinearity (saturation), it is much more likely to erase magnetic tape than a normal power transformer. True; False.
6. Although admittedly primitive, measuring power feed line filters in situ is acceptable for variant loads. True; False.
7. If the bandwidth of a low-pass filter is equal to the reciprocal of pulse width, a square wave is changed into a sinusoidal waveform. True; False.
8. Far-field measurements of RFI sources, according to VDE standards, are made at a 30 m distance. But the standards do not stipulate providing a metal grid ground plane (which, however, is tacitly assumed and in fact hidden). Does the presence or nonpresence of the ground plane make a great difference in the test results? Yes; No.
9. For hand tools having plastic housing, VDE standards allow two wiring options for mandatory filtering: (a) three wires, or (b) two wires only, but double insulation. Which is easier to filter? (a); (b).
10. A scope has a bandwidth of 10 MHz. Is a signal with a 100 ns rise time faithfully displayed on it? Yes; No.
11. The best low-pass filter for a low-impedance noise source and a high-impedance load is (a) a choke; (b) an LC filter; (c) a shunt C.
12. Can common mode/normal mode conversion occur in a balanced (symmetric) line? Yes; No.
13. Does the green wire improve EMC? Yes; No.
14. If a cow had brains, in a thunderstorm would she stand under a tree with her head toward the trunk? Yes; No.
15. The lowest resonance frequency of a 15 m line shorted at both ends is (a) 10 MHz; (b) 20 MHz; (c) 30 MHz.
16. It is possible to extract a weak, well-defined signal out of overwhelming white noise, although filters do not work at all. Yes; No.

17. It is not advisable to connect the shield of the main power cable to the building ground or steel. Yes; No.
18. Should the green wire be fed through the core of the ground fault interruptor? Yes; No.
19. Should most normative standards be kept as they are? Yes; No.
20. An abnormal transient can be increased by increasing C. True; False.
21. Given the same amplitude, a short pulse is more likely to cause electrical breakdown than a long pulse. True; False.
22. Does the common mode rejection of isolation amplifiers decline with increasing frequency? Yes; No.
23. What kind of resistor is much better for use in spike suppressor RC networks? (a) carbon composition; (b) carbon film.
24. One winds a signal line pair around a ferrite core to absorb common mode energy. True; False.
25. A power feed line filter working essentially in short circuits at both sides will never show an insertion gain. True; False.
26. In a quasistatic electric field, is a thick metallic shield better than a thin one? Yes; No.
27. In Problem 26, replace the metallic shield by a high-ε shield. Is your answer still the same? Yes; No.
28. Mounted in a metallic shield, a metal pipe of 10 cm diameter will not pass frequencies lower than 1.76 GHz. True; False.
29. In Problem 28, does this still hold if a well-insulated wire is passed through the tube very close to the wall? True; False.
30. Should limitative standards be reviewed every few years? Yes; No.
31. Initially, a capacitor acts like a short circuit if a step voltage is applied. True; False.
32. Hand tools used for manufacturing and repair are a serious source of EMI. True; False.
33. Which two critical parameters are the key to good magnetostatic shields? (a) size of shield; (b) thickness of shield; (c) permeability; (d) dielectric constant; (e) conductivity.

Problems: Group A

34. What is safer for reduced step voltage? (a) underground guarding [Table 9.2(d)] or (b) isolated ground rod according to Figure 9.1(b).
35. In a normal switching transient the switched amplitude may maximally be multiplied by (a) 2; (b) >10.
36. What FATTMESS parameter is the key to the great advantage of fiber optics for EMC? (a) T; (b) M; (c) E.
37. The desideratum for a guard shield of an amplifier is to make sure the common mode current flows into the guard such that the common mode/normal mode conversion is minimized. Yes; No.
38. The safety wire is (a) black; (b) white; (c) green.
39. Twisted wires prevent common mode induction. Yes; No.
40. The disadvantage of a V/F converter is that it cannot be synchronized by the system clock. True; False.
41. A shield in a transformer reduces the interwinding capacity, thus reducing common mode transfer. True; False.
42. The shield just discussed should be closed. Yes; No.
43. Should a filter for ac lines be designed for both normal mode and common mode operation? Yes; No.
44. A snubber, having too large an RC, may slow down the action of a relay prohibitively. Yes; No.
45. What material do you use for good electrical insulation but good heat conduction? (a) Teflon; (b) beryllia; (c) aluminum oxide; (d) mica.
46. Under what conditions is the use of a large filter capacitor "verboten?" (a) to minimize 60 Hz leakage currents of line filters; (b) to minimize common mode/normal mode conversion in a balanced line driven unsymmetrically.
47. A feed-through capacitor is needed for good high-frequency performance of a filter. True; False.
48. It is always best to have the lowest possible clock frequency. Right; Wrong.
49. A steel tubing (B_s = 10,000 G, 1 mm wall thickness) is subjected to an external field of 100 G. Mark the diameter of

of the tube for which it becomes saturated. (a) 5 cm; (b) 10 cm; (c) 20 cm.

50. For radiated interference, double grounding is immaterial. True; False.

12.3.2 Answers: Group A

1. Far field
2. True
3. No
4. No
5. True
6. False
7. True
8. Yes
9. (b)
10. Yes
11. (b)
12. No
13. No
14. No
15. (a)
16. Yes
17. Yes
18. Yes
19. Yes
20. False
21. False
22. Yes
23. (a)
24. True
25. True
26. No
27. No
28. True
29. False
30. Yes
31. True
32. True
33. (a) and (c)
34. (a)
35. (a)
36. (b)
37. Yes
38. (c)
39. No
40. True
41. True
42. No
43. Yes
44. Yes
45. (b) and (c)
46. (a) and (b)
47. True
48. Right
49. (b)
50. False

12.3.3 Problems: Group B

51. What is the purpose of modulation for LED sources?
52. Two toroids of equal size are wound by a common winding constituting 1 µH in air. At 150 kHz the toroids have a permeability of 100 and 100 - j100, respectively. What is the impedance of the arrangement?
53. How is the display of a spectrum analyzer affected by the pulse repetition rate (prr)?
54. Why is the use of an absorption clamp (loading line with several hundred ohm via ferrite toroids)--as used in Europe for measuring conducted interference current--to be taken with caution?
55. A ceramic pi filter consisting of two capacitors and a ferrite bead is affected by bias and temperature such that the capacity is reduced to 1/3 and the ferrite impedance to 1/2. How much is the insertion loss reduced?
56. Which two simple checks would you do to ascertain that a power feed line filter would work satisfactorily under changing interface conditions that you cannot predict?
57. A piece of wire--with a length/radius ratio of 40:1--sticks out of a metal plate on which a field strength of 100 V/cm exists. What is going to happen?
58. How and how much is a 100 µs 10 V notch reduced by a low-pass filter having a sharp cutoff frequency of 1 kHz?
59. Why should you observe the shape of the envelope of the beat frequency when making frequency comparisons?
60. Why is it absolutely necessary to provide filters on each lead of a filtered connector?
61. Why can the transfer impedance of a tubular ceramic feedthrough capacitor not be smaller than the characteristic impedance of the equivalent transmission line--usually on the order of magnitude of 0.5 ohm?
62. A 100 A rectangular pulse runs along the outer brass conductor of a solid coaxial line. The thickness of the brass is 1 mm.

What pulse width cannot be exceeded if the peak of the current induced on the inside conductor must be less than 10 mA (80 dB). Is it 10, 1, or 0.1 μs?

63. A circular waveguide of 0.5 cm radius and 2 cm length is practically filled with fiber-optic cables constituting an effective ε of 6. What is the cutoff frequency of the "filled" waveguide?

64. Looking at the responses both in the frequency and time domains, a shield can be compared with what kind of filter?

65. When is it better to ground only one end of a cable? When is it better to ground both ends? When should a pigtail connection be replaced by a backshell?

66. What happens to self-resonances in filters under the influence of current or voltage bias?

67. How many decibels are in 1 neper?

68. On what side of a transformer do you best place an MOV?

69. List the advantages and disadvantages of electrooptical isolators?

70. Junction rectification is a cause of semiconductor malfunction. Why is it less critical at very high frequencies?

71. High-resolution A/D converters are very noise sensitive. Calculate the peak-to-peak voltage of the LSB of an A/D converter (±2 V, 12 bits).

72. Calculate the skin depth of soil, with a resistivity of 1000 ohm-m, at 1 MHz. Use a nomograph for Cu, which has a resistivity of 1.72×10^{-8} ohm-m, to simplify your calculations.

73. Roughly, what is the field strength, electrical and magnetic, 1 km away from a 25 kW transmitter?

74. From what frequency on can you expect deep dips in the shielding attenuation for a chicken wire cage, being a cube of 3 m dimensions?

75. What is the self-resonance of a printed circuit board for which the following data are given: 30 cm long, dielectric constant of 4, one side metallized, ground plane is shorted at one small edge via capacitors?

Problems: Group B

76. A lightning strike of 10 kA/µs rise time and 100 kA peak value hits a lightning rod that leads over a wire or 10 µH to a ground resistance of 10 ohm. Water pipes, grounded elsewhere, come as close as 1 m to the lightning conductor. Taking into account a field enhancement factor of 5 (due to field concentration close to the wire), could you expect a side flash? Proof it.

77. Figure 12.1 presents the attenuation of a 1 m ignition cable for three different structures: a cable with a 40 µH inductor (+30 ohm); a lossy cable (wire spiral on flexible ferrite and plastic core (Bougicord developed by LEAD); a resistive cable (10 kohm). Identify which (a), (b), and (c) are and compare their pros and cons.

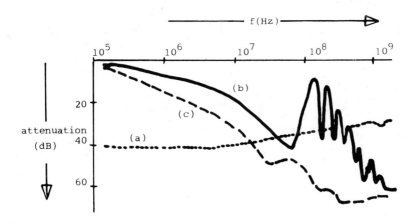

Figure 12.1

78. Figure 12.2 sketches the use of an absorption clamp, widely used in Europe to measure conducted interference. State the purpose of (a), (b), and (c) and the limitations of the device.

79. State the causes of and remedy for the various kinds of interference shown in Figure 12.3.

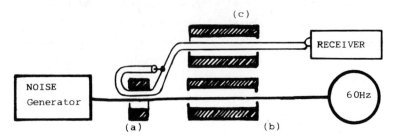

Figure 12.2

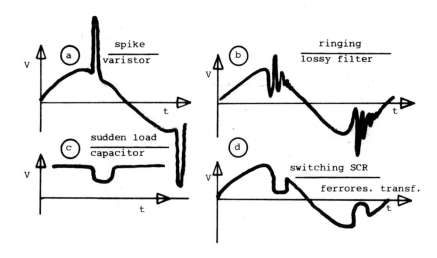

Figure 12.3

12.3.4 Answers: Group B

51. Immunity to ambient light.
52. 50 ohm + j100 ohm.
53. If the prr is smaller than the resolution bandwidth: Fourier transform; repetition rate line display; narrow-band amplifier gives transient response. If the prr is large: Fourier series; narrow-band amplifier gives CW response.
54. Represents only high-impedance load.

Problems: Group B

55. 31 dB.
56. Select whether multisection filter and check for ringing.
57. Arcing, because field strength is 45 kV/cm on wire end.
58. 2 V.
59. Entrainment: wrong difference frequency.
60. Filters are bypassed by inductive coupling.
61. $V_2/I_1\big|_{I_2=0} = 1/a_{21} = -jZ_0/\sin b$ (= Z_0 for b = 90°) (see Table 2.1 for derivation).
62. 1 µs.
63. 14 GHz.
64. Low-pass filter.
65. Low frequencies; high frequencies; high frequencies.
66. Shift to higher frequencies.
67. 8.7 dB.
68. Side of spike source--utilize leakage reactance of transformer.
69. Advantages: no ground loops, common mode suppression, safety. Disadvantages: limited linearity and range, temperature sensitivity, limited speed, cost.
70. Capacitive shunt.
71. 2^{13} = 8192: 2 V/8192 = 0.25 mV.
72. 0.066 mm $(10^3 \times 10^8/1.72)^{0.5}$ = 16 m.
73. 1.5 V/m; 4 mA/m.
74. Since λ equals 3 m (100 MHz), first dip at 50 MHz.
75. 125 MHz.
76. 10^6 V + $j10^5$ V; 5 MV/m much larger than 3 MV/m, which is breakdown voltage of air. Hence the answer is yes.
77. (a) Resistive cable: attenuates to much low frequencies (energy!); (b) with inductor: pronounced resonances, no good; (c) lossy line: best solution.
78. (a) Current transformer; (b) simulating high load impedance; (c) prevents surface currents on cable surface. Measures interference currents for high load impedance at high frequencies only, hence is of limited use.

79. (a) Transient spike--use voltage limiter; (b) ringing filter--use filter lossy in upper passband; (c) sudden load--use capacitor across line, active filter or regulator; (d) notching by SCR--use ferroresonant transformer or more expensive regulator.

12.3.5 Problems: Group C

80. The double exponential--as a response of a resonant circuit to a step function (Chapter 1)--starts suddenly with a maximum slope. But the actual step response starts with a smooth transition from the unswitched state. Explain.

81. At the input of a dc/dc converter the following measurements were made: 45 V at 10 A and 15 V at 30 A; an LC filter (L = 100 µH, C = 500 µF tantalytic with 0.1 ohm series resistance). What happens? How can you prevent the trouble?

82. The eleventh harmonic of 60 Hz, amplitude 100 V, is feeding a 4.5 m antenna having a capacitance of 25 pF. What power is radiated?

83. An oscillator is driven as follows: f_0 = 1 MHz; Q (of tank) = 50; open-circuit amplification at f_0 given by $V_0 = 110(V_i - 0.1\ V_i^2)$. What is the range of synchronization if 0.5 V of 500 kHz is coupled into amplifier input (V_i)? Remember: $\sin^2 x = (1 - \cos 2x)/2$.

84. Make a real effort with this problem of A/F conversion, because understanding its full implications will greatly benefit your EMI hardening ability: A/F-converted analog signals and digital signals have the great EMC advantage of fixed high-amplitude transmission such that low values of the original analog signal do not disappear in noise. This is particularly significant in high-resolution data spans. In view of the great merits of A/F conversion, elaborate more on it in conjunction with:

(a) Sensors (e.g., what kind of tachometer is best, what not?).

(b) Comparing AM with FM, show that the low values of a range

of analog signals are even more noise hardened than the upper part of the scale.
(c) Amplifier stability.
(d) Impulsive noise in digital data trains.
(e) The present limitations of fiber optics (linearity, etc.).

85. A cable is grounded at both ends. Where would you place ferrite toroids on the cable to suppress (a) the lowest resonance frequency; (b) the next higher resonance frequency?

86. A practical grounding of whip antennas on metallic vehicles--only insignificantly impairing reception--is as follows: The whip is directly connected to the chassis, where a ferrite bead is placed on the whip. The antenna circumflexes the bead with the smallest possible loop, with the inner conductor connected to the grounding point and the outer conductor going through and connected to the metallic base. Explain how it works.

87. What is wrong in Example 8.3?

88. A certain MOV material is formed into a tubular feed-through capacitor with an electrode length of 0.5 cm. The insertion loss of an RC filter, with this capacitor, shows a pronounced dip at 500 MHz if measured with small voltages. What is the dielectric constant of the material? Any idea how to exploit this?

89. For a single-phase clock, master-slave flip-flops are characterized by a shorter time t_1 during which all registers are opened and loaded and a longer time t_2 during which the combinatorial networks act and where the inputs are closed. For what purpose can you exploit this industry standard?

90. A 100 ns pulse, inverted, rides on 10 V, 60 Hz. Without falsifying the 60 Hz sine wave, to what lowest value can the pulse be reduced with a slew rate filter.

91. Two configurations of ceramic feed-through capacitors are given: (a) straight tubular, C = 2000 pF; (b) same, but the outer electrode split radially in the middle, but all around

reconnected again over a ferrite bead constituting 20 ohm at VHF frequencies. What is the transfer impedance of both structures at 150 MHz? Assume no resonances and remember that for a pi network $Z_{21} = a_{21}^{-1} = Z_1^2/(Z_2 + 2Z_1)$, where Z_1 is one of the two equal shunt impedances, Z_2 the series impedance of the pi. Explain why a negative transfer impedance--if you should find it so--is not a mistake.

92. A short line, terminated at each end by 20 kohm, is coupled, by a distributed capacitance of 10,000 pF, to another line by which a load is switched on a 100 V source. What is the response in the first line in terms of V(t) and energy?

93. Figure 12.4(a)-(c) refreshes your memory as to what happens to a steep wave front traveling along a line when it meets a discontinuity. Now consider (d): What happens initially to the step voltage arriving at point P? $Z_0 = 25$ ohm, $C = 0.5$ µF.

94. Two shielded structures, wired as shown in Figure 12.5, are hit by a lightning strike of 30 kA peak and a rise time of 3 µs. What are the peak currents at points B and C? What are the voltages between points AB and BD?

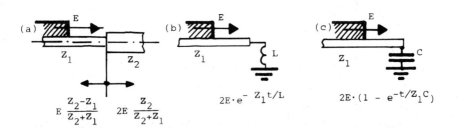

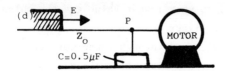

Figure 12.4

Problems: Group C

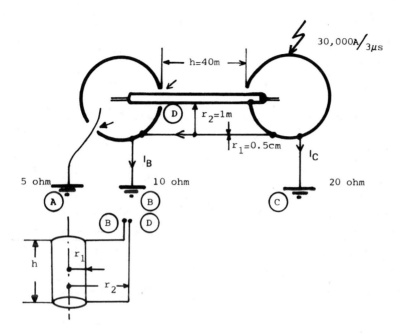

Figure 12.5

95. To avoid cross-current conduction in a push-pull dc/dc converter, the two driven bases of the switching transistors are preceded by RC filters where the R is shunted by a diode.
    (a) Sketch the circuit and explain the cross-current.
    (b) Name the three major detrimental effects of cross-current conduction. (c) How do the remedies described before work?
96. A 1000 V step is fed into a series resonant circuit. Z_0 = 50 ohm, Q = 20, source impedance zero. Sketch the current flowing.
97. For a filter inductor you need a low Q for the range 5 to 15 kHz. You chose to exploit heavy laminations (eddy current losses). But you are unhappy that, at higher frequencies, the inductive component increases only with the square root of frequency. What simple thing can you do?
98. Sketch arrangement described in Problem 86. Explain why it will not work as protection for transmitter antennas.

99. In braided coaxial cables the transfer impedance starts to increase with frequency from about 1 to 10 MHz, a very undesirable phenomenon. Use your imagination as to how you could approximate the effect of a solid cable, but still using braids.

100. Two shielded enclosures are grounded separately such that a 60 Hz voltage difference exists between them. (They are quite a distance apart.) A strong RF field necessitates continuous shielding of the cables connecting the two structures. Most critical are the lines leading from sensors in the left structure to data processing equipment located in the right structure. Sketch and explain several ways to overcome the ensuing problem.

101. *Strategic business planning for EMI hardening.* Marshal the key points needed for a convincing presentation to management (of a small to medium-sized company) concerning cost-effective EMI hardening of a planned new industrial control system. Hereby you must state:

    1. The *need* for EMI hardening
        A. By delineating the short- and long-range consequences (down-time, hazards) occurring at the customer's plant (see Chapters 5 and 11)
        B. By pointing out the consequential negative reflections imparted by item A on your own business (law suits, loss of future business, etc.)
        C. By stressing the clear competitive edge you gain by appropriate hardening
        D. By quantifying the interference limits the system will safely withstand
    2. The crucial *constraints* under which these needs must be met:
        A. Uncertainty of present and unpredictability of future levels of interference at the plants of prospective customers

Problems: Group C   267

   B. Lack of standards for industrial EMC
   C. The competition provides some kind of hardening
   D. The customer will not pay a high price for hardening
   E. The company cannot afford the very expensive test facilities and instrumentation normally needed for EMC
   F. For the foreseeable future only the domestic market is of interest

   Most important:

3. You must, by using your imagination, arrive at a good *solution*: Given the contrary needs and constraints described, outline a strategy of attack to minimax costs versus benefits, both for the company and the customers. Based on Chapters 5 and 11, preplan two key points: economic hardening of system and economic EMC testing.

12.3.6 Answers: Group C

80. The double exponential is based on one L and one C. But the L has a distributed C and the C has an unavoidable lead L. Thus, e.g., in a series resonant circuit, the L is initially shunted by a C, although for a very short time only.

81. $[(45 \text{ V} - 15 \text{ V})/(10 \text{ A} - 30 \text{ A})] = -1.5$ ohm, which is in parallel with $(L/CR_t) = 100 \times 10^{-6}/500 \times 10^{-6} \times 0.1 = 2$ ohm. Oscillation can be suppressed by lowering the L/C ratio.

82. With $(h/\lambda) = 10^{-5}$ the radiation resistance is 1600×10^{-10} ohm $= 1.6 \times 10^{-7}$ ohm. The current, determined by C, is given by $100 \text{ V}/104{,}000 \text{ ohm} = 1$ mA, and $I^2 R = 1.6 \times 10^{-13}$ W.

83. $M = (110 \text{ V}/100 \text{ V})[\ldots - 0.1 \times 0.5^2 \times \sin^2(2\pi \cdot 500 \text{ kHz})] = 0.0138$ V. Hence $M/S = 0.0138 = 2\Delta f_o Q/f_o$ or $\Delta f_o = \pm 138$ Hz.

84. (a) Do not use the customary electrical generator plus rectifier as velocity sensor (low speed means low voltage). Rather, use a light-chopping disk to drive a light-sensitive diode providing pulses the repetition rate of which is proportional to the velocity.

(b) For instance: For FM we have a modulating frequency of f_m = 10 to 10,000 Hz and a peak frequency deviation of, let us say, F = 40 kHz. Then the noise reduction, in terms of AM, is the reciprocal of the deviation index D = F/f_m. Thus: D = 4000 to 4. Now, similarly for our converted signal: If f_m = 0.1 to 100 Hz, and F = 400 Hz, we arrive at the same D = 4000 to 4. That means that the low signal levels are much more noise immune than the high signal levels, due to conversion.

(c) Drift and other amplifier effects are negated.

(d) In A/F conversion we can count and arrive at a minimal error, whereas in digital transmission a significant bit may be falsified.

(e) We can use cheap electrooptical isolators to open ground loops without losing accuracy.

85. (a) At ends; (b) in center also.
86. A lightning strike or other strong current saturates the ferrite such that during the disturbance the receiver is protected.
87. Multiplier 3 refers to polar concentration (Table 1.4). Replace 3 by 4 (see Fig. 1.11). There is no significant change in the end result.
88. Epsilon is 3600: a very economic combination of filter and limiter.
89. Do disturbing switching during t_2.
90. $(d \times 10 \text{ V} \times \sin 377t/dt)_{min}$ = 3770 V/s = I_p/C must be met to avoid disturbing the 60 Hz. Then spike can be reduced to 377 μV.
91. (a) -j0.5 ohm; (b) Z_1 = -j ohm, Z_2 = 20 ohm. $(1/a_{21}) \approx (-)1/20$. The minus sign is not wrong; a fictitious series element must also be considered such that as seen from accessible terminals no negative resistance exists.
92. Exp function, starting with 100 V, decaying with a time constant of 0.1 ms. Neglecting 0.36^2 against 1, the energy turns out to be about 50 μJ.
93. Initially (b) is dominant; hence 2E decaying with time constant of 12.5 μs.

94. 20 kA at B, 10 kA at C. AB: 200 kV; BD: 280 kV. Concerning V_{BD}: $V = -N\, d\phi/dt$; $\phi = \int B\, dA = \int^h \int_{r_1}^{r_2} (\mu I/2\pi r)/dl\, dr$, and with $\mu = 4\pi \cdot 10^{-7}$ H/m, $\phi = 2 \times 10^{-7}\, I \times h \times \ln(r_2/r_1)$ (Wb) or $d\phi/dt = 2 \times 40 \times 5.3 \times 10^{-7} \times (2/3)10^{10} = 280$ kV, with $(2/3)10^{10}$ being dI/dt.

95. See Figure 12.6 for (a). (b) EMI, efficiency, reliability. (c) RC for delayed start, diode for fast turn off.

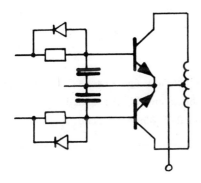

Figure 12.6

96. See Figure 12.7.
97. Add, under same winding, smaller height of core having fine laminations. Result: See Figure 12.8.

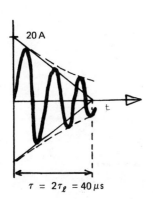

Figure 12.7

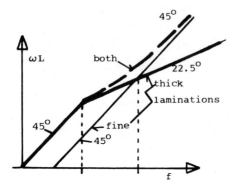

Figure 12.8

98. See Figure 12.9. Ferrite heats up fast and cracks.
99. See Figure 12.10. As material between the two braids, use a mixture of rubber and ferrite powder (done by LEAD).
100. See Figure 12.11

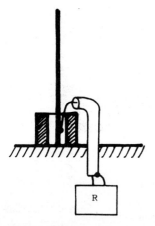

Figure 12.9

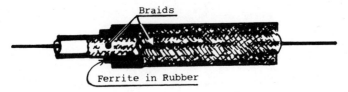

Figure 12.10

Problems: Group C

99. See Figure 12.10. As material between the two braids, use a mixture of rubber and ferrite powder.
100. See Figure 12.11.

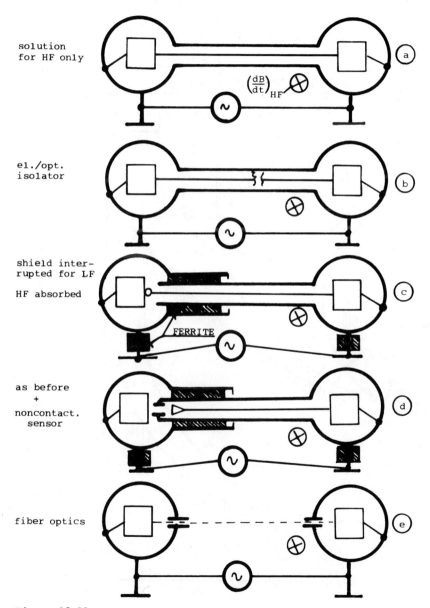

Figure 12.11

12. Problems and Solutions

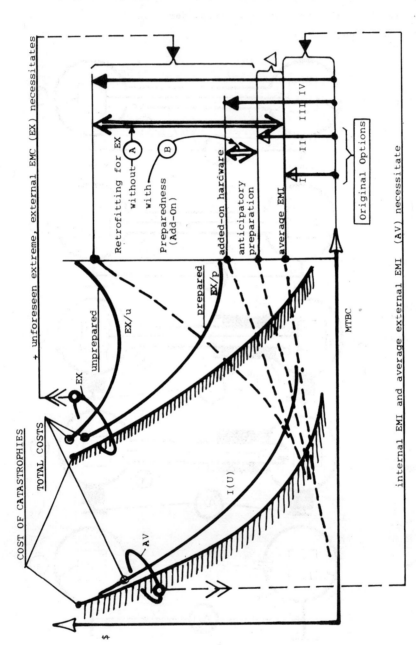

Figure 12.12 Anticipatory Cost Reduction

Problems: Group C

101. The following two cornerstones of the plan are suggested:
    1. *Two-option approach* (Fig. 12.12) The customer is given the choice of two options:
        A. Option I(U) for internal EMI and average external EMI, caused by average (AV) EMI, but Unprepared for extraordinary external EMI (EX), as it could happen, e.g., if somebody would move close with very noisy equipment.
        B. Option II(P), as I(U), but in addition, for a relatively small cost Δ (see Fig. 12.12 with anticipatory Preparation for easy add-on EMC measures in case of unforeseen, extraordinary external EMI.

    The great advantage of II(P) over I(U) for the customer should be emphasized. There is often no legal recourse if somebody erects close by a facility that creates strong interference. In case of EX, sudden high external interference, the cost of retrofitting option I(U) for the changed conditions is very high, namely A = IV - I. In the case of option II(P), the cost of retrofitting the system for the much stronger, originally not present EMI is low, namely only B = III - II, because there is no need to disassemble and reassemble completely large portions of the system for shielding, grounding, and filtering, because the system is prepared by foresight. Naturally, we have to identify the add-on hardware and its installation in the system manual.

    2. *Fitting EMC testing* (Fig. 12.13) Beyond the contingency planning just described, we must drastically reduce the testing costs. Conventional EMC testing can run into several hundred thousand or even millions of dollars. Thus:
        A. Instead of buying an expensive large, shielded room with 100 dB attenuation, we build an inexpensive one of fine-meshed chicken wire. We filter it and make it large enough to house our system. Thus we avoid

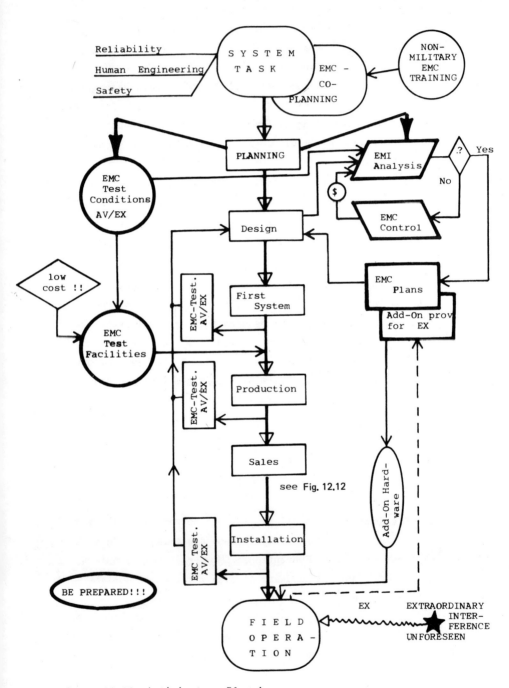

Figure 12.13 Anticipatory Planning

getting in trouble with the FCC or creating trouble for our computers or other sensitive equipment.

B. Instead of using very expensive automated, computerized swept-frequency test systems (containing broadband power amplifiers, spectrum analyzers, computer printout, etc.), we apply:

(1) For AV conditions:
  (a) Run the system with all critical ICs at 70°C. Does it work well if a few cycles of 60 Hz are skipped? (Effect of operation of lightning protector.)
  (b) Put impulse generator at 60 Hz entrance of each subsystem.
  (c) Touch the housing of each subsystem with an ungrounded, running hand drill (make sure you do not get a shock yourself).

(2) For EX conditions, apply to each subsystem and critical information line:
  (a) The relay noise test (Chapter 11).
  (b) The motor noise test (Chapter 11).
  (c) The antenna of a radio amateur transmitter (different orientation, up to 1 m proximity). Consider other frequencies.
  (d) In addition, we may use the impulse generator to feed a Lecher line to be coupled into critical information links. Operate the line with 0.1, 1, and 10x Z_0 load.
  (e) Self-interrupting 110 V, 60 Hz, via 100 ohm, between enclosure and system ground.

The system must work under all conditions above, including 70°C.

C. For reproducibility and for incorporation into the systems manual and sales brochure, we measure:

(1) The peak electrical field strength $E_p = V_p/m$ by using a peak-reading voltmeter or storage scope.

(2) The maximum (dE/dt) from reading the storage scope.

(3) The maximum (dB/dt), by using an electrostatically shielded loop [$V_m = L(di/dt)_m$] and a storage scope.

(4) For high-frequency sensitivity, determine the field strength from the power and distance of the sender.

Thus we determine our test data backward, so to speak: We increase the stress until the system fails (catastrophy). Or if we cannot reasonably create a catastrophy (by reaching the limit of our interference sources), we still can state a safe limit of stress constituting quantifiable worst-case conditions.

If, in specific cases, we whould have to meet such standards as Mil. Std. 461/2/3, CISPR, or VDE standards, requiring expensive test gear, we can always farm out the testing to a reliable testing firm or rent test equipment.

D. See also Section 12.4.5 (FCC regulations) for emission testing.

12.4 COMING ON STREAM, MAYBE

Industrial EMC is still in flux. Hence, as we come now to the end of the book, we shall comment on some still somehow embryonic (or at least incomplete) developments which in all likelihood will significantly affect EMI control. Let us outline and assess these specific, promising trends in the context of those (control) chapters into which they belong.

12.4.1 Inductor-Less Voltage Limiters

In Chapter 5 (localized suppression) we encountered the detrimental secondary effects of fast-switching limiters: Their not-zero L across the line causes spikes of great amplitude. Many scopes are not fast enough to see them but they are there nevertheless.

Coming on Stream, Maybe 277

We can eliminate such destructive spikes by using feed-through geometries (no-L) as is done with ceramic feed-through capacitors. This is particularly easy in the case of MOVs. But because of their high dielectric constant (several thousand), parallel resonances occur, for tubular structures, in the VHF-UHF range (Problem 88). Resonances can be shifted to much higher frequencies (where they do not disturb) by using discoidal structures or cross-slotting (deresonating) of tubular structures (U.S. Patent 3,007,121). No truly no-L voltage limiters are yet on the market.

12.4.2 Some Specificities on Shielding

Pertaining to shielding, three developments are of particular interest:

1. *Earthen shielded enclosures.* In Chapter 2, for the sake of simplicity, we calculated the shielding effectiveness of holes in the ground on the basis of spherical coordinates, applied to Maxwell's equations. The results are equally valid for prolate spheroids, the focal points of which can more conveniently be the loci for sender and "victim." This "dirt-cheap" solution, significantly improved by bi-jagged, nonlossy (fireproof) transition cones of porous brick (structural support), looks quite promising in closely approximating undisturbed free-space conditions in a very economical way.

2. *Very low transfer impedance cable.* In Chapter 2 we also discussed the poor transfer impedance of braided cables above 1 MHz, and in Problem 99 we saw a simple solution to improve this situation. F. Mayer, of LEAD, France, has developed a multiple-braid cable as follows: This coax cable contains three braids and two interlayers of Musorb (exploiting the dielectric and magnetic losses of ferrites embedded in plastics or rubber). The z_t of such a cable is on the order of microohm/m (at very high frequencies). Hence, for a length of 1 m, we can expect only 1 μV disturbance for a current of 1 A on the outer braid. Such cables are much more economical than

solid metal shielding conduit. And they are much more flexible, even if the conduit is convoluted. It is anticipated that such cables will be in use soon, once sufficiently good connectors and backshells, commensurate with the excellent transfer impedance, are available. One must always keep in mind that a very good cable is of no use if the connector is not really tight electrically. Quite often, tightening a connector with a wrench instead of by hand makes a great difference for frequencies above the 10 MHz range. Improved connectors should preferably contain ferrite to absorb leakage at the end of the cable.

3. *Shielding of plastic housings*. There is a pronounced trend to replace metallic housings or cabinets for electrical equipment, in particular consumer goods, by plastic ones. The reasons are: lower price and weight, and greatly improved design flexibility, resulting in better eye appeal and thus salability of the end product. But plastic enclosures are pervious to EMI, and many designers forget this. Viewing windows are available with embedded wire mesh. The plastic housing itself can be made electrically shielded by two generic methods: conductive coating or conductive plastic.

   A. *Conductive coating* can be provided in quite a variety of ways, to be decided upon by the number of pieces to be shielded and by their complexity: metal foil or tape, ion plating, spray plating, vacuum metallizing, cathode spraying, electroless plating, flame spraying. Most metallic paints you can buy in paint stores are not conductive, hence worthless.

   B. *Conductive plastics* (composite materials containing metal flakes or metallized fibers) are more economical and still provide sufficient shielding (40 dB or so). There is, however, one drawback at present: The surface layer of such composite materials is predominantly plastic, hence an insulator. This impediment, unless properly corrected by molding-in metal inserts or grinding off the non-conductive surface layer, has two very bad effects:

(1) No contacts where contacts are needed for shield continuity (joints, shielded windows, etc.).

(2) The high surface resistivity does not bleed off static electricity, which can kill transistors or solid state memories. Sometimes the addition of carbon black reduces the surface resistivity sufficiently. In Europe, the surface resistivity is mandated to be less than 10^9 ohm per square. No corresponding standard exists in the United States.

12.4.3 Trends in EMI Filtering

Controversies still permeate interference filtering, so much so that one can speak of filtering as the most disreputable and dissatisfying area of EMC. Quite often, hanging onto convenient, but outdated concepts are responsible for this lamentable situation, which is slowly being overcome by reason and insight.

In this context the reader should be aware of the particular roles played by intentional losses introduced into various EMI filter classes:

1. In lumped-element filters, losses must be provided in the transition band only to prevent ringing and/or insertion gain. Such filters reflect the interference.
2. In contrast, for distributed element filtering lines, we exploit frequency-dependent losses for the stopband (RC filters). The interference is absorbed. By their very extension such lossy cables can dissipate large amounts of power with ease.
3. The situation is very different for cascaded (split-electrode) ceramic filters utilizing ferrites for a series element that is predominantly resistive (absorptive) at very high frequencies. The compact structure of such lossy ceramic filters can absorb only limited power (a few watts). Otherwise, the ferrite cracks and renders the filter useless.

Lumped-Element Power Feed Line Filters

(For background, see Fig. 1.12, all of Chapter 8, Secs. 11.1 and 12.2, and also Problem 97.) Reluctantly, and without realizing their potential market gain, some filter manufacturers are beginning to develop nonringing filters (losses in the transition band), as it now seems quite certain that eventually standards for testing worst-case filters will be issued by CISPR and ANSI. Unfortunately, the term "worst case" somehow implies for many an engineer that it occurs only seldom. In this respect "worst case" is an unfortunate misnomer not adequately describing the actual situation, which is characterized by unavoidably indeterminate and variant interface conditions encountered in large control systems, where the "worst case" is quite normal.

Distributed Lossy EMI Filters (Applicable to Both Normal Mode and Common Mode)

Lumped-element filters (see above) are the only answer if we need to filter below 1 MHz (or so). If, however, we have to filter well above 1 MHz, we may soon have the economic option of French-developed lossy-line filters (F. Mayer, LEAD) using the same Musorb material already discussed in conjunction with low-z_t cables (Sec. 12.4.2). This material extruded onto wires yields truly lossy lines, the more so the longer the line and the closer the line is to ground. (Attenuation due to series element is proportional to $R/2Z_0$.) For best results, one can coat the phase and neutral wires with Musorb and replace the green wire by a braid about the other wires. Then a 5 to 10 m line length will attenuate more than 50 dB above, let us say, 30 MHz. And rise times of voltage spikes will be greatly reduced. Moreover, adding relatively small feed-through capacitors (say 0.1 µF) will bring the good filtering performance to much lower frequencies or cable lengths. Such truly lossy lines are presently being introduced in Europe and may be expected on the U.S. market in the foreseeable future. UL approval of such filtering power feed cables will be necessary. Watch for passband resonances.

Some Comments on Ceramic Feed-Through Filters

We discussed certain critical aspects of ceramic filters in conjunction with Table 1.9. See also Problems 55, 61, 88, and 91. And we did not go into any further details, as manufacturers' catalogs are usually rather specific. The author's patents [2] describe various structures and principles of ceramic filters. Not all of them are mass produced.

Attempts are being made at some places to co-fire the dielectric and the ferrite in order to reduce manufacturing costs and to improve the mechanical strength. But in spite of the application of supposedly protective layers, some ferrite material diffusing into the dielectric cannot always be fully prevented. This results in a reduction of the breakdown strength. Thus when using such ruggedized filters, one should check carefully for electrical reliability.

12.4.4 The Grounding Dilemma: Another Possibility of Solving It

(For background, see Tables 1.7 and 2.3, all of Chapter 5, Sec. 9.3, and Problem 94.) Double grounding of cables poses a problem that makes many an engineer ill at ease: At low frequencies, single grounding is required to avoid leakage current. But at high frequencies, the need for hole-free through-connections to the grounded housings at both ends of the cable is a contrary requirement. As long as we cannot apply fiber optics without hesitation (costs, aging, limited dynamic range, etc.), the solutions given in Problem 100 would merit realization if other means, e.g., triax cables, of isolation (Chapter 5) are not advisable.

12.4.5 FCC Regulations Regarding EMI Control of Digital Equipment

Throughout the book, in particular in Section 1.3 and Chapters 5 and 10, we stressed the limitations and frustrations imparted by imperfect, unrealistic, and contradictory standards. A new FCC regulation [3,4] is, but must not be, upsetting the industrial control community. It imposes clearly defined limits on conducted and radiated emission (up to 200 MHz) from digital equipment and describes how to measure it. The rationale behind this standard is to prevent harmful interference in communications equipment.

The FCC mandates that the standard be met by self-certification, with the FCC reserving the right to spot-check. The equipment maker must not only provide a corresponding label on the equipment, but also must put a pertinent comment into its manual. The system designer can do the testing with loaned test equipment or can have the test done by an EMC testing company. At any rate, as the designer has to EMI-harden the system anyway, the normally-not-made test for emission may be considered as a test complementary to susceptibility (as the required isolation works both ways).

REFERENCES

1. Proceedings of the 1978 EMC Workshop, National Bureau of Standards, Special Publication No. 551.
2. U.S. patent Nos. 2,973,490; 2,983,855; 2,994,048; 3,007,121; 3,023,383; 3,035,237; 3,243,738; 3,329,911.
3. Federal Register, Vol. 45, No. 123, June 24, 1980, Proposed Rules, pp. 42347-42358: FCC/47CFR, Parts 2 and 15 (General Docket No. 80-284, FCC 80-335): Verification and Methods of Measurement of Computing Devices.
4. FCC Controls Digital Equipment, A. Wall and R. Bromery, International IEEE EMC Symposium Record, 1980.

Index

Absorption:
  clamp, 260
  by soil, 80-84, 277
  (see also Shielding)
Acronym (FATTMESS), 32, 47
A/D converter, 258
Air:
  adsorption, 23
  gaps, 21, 22
All-pass, 73
Alumina, 22
Aluminum:
  pinholes, 41
  tunneling, 36
  wiring, 22, 35
Amplifier, 75
a_{mn}, 207-214
Amplitude (see also
  Nonlinearity), 48
Antenna:
  transmitter, 265
  very short, 77
Architects, 232
Arrest, respiratory, 101
Arrythmia, 101
Asphyxia, 101
Attenuation:
  by buildings, 76

[Attenuation]
  vs. insertion loss, 203
  measure, 53, 61
Averaging, 159

Backshells, 66
Bandwidth, 71, 253
Berryllium oxide, 22
Bias, 49, 50, 184, 223
Blanking, 153
Bonding, 50, 54, 128, 131, 135, 229
Bougicord, 259
Braids, 66
Breakdown strength, 21
  of air, 18
Breakdown voltage:
  of ceramic capacitors, 23
  secondary, 93
  in semiconductors, 34, 93
Buffer, 196
Building:
  electrical problems, 165
  highrise, 136
  and lightning, 126
  pointers for architects
    concerning, 232

Bureaucracy, 40
Business planning, 266

Campbell, 196
Cancelling, 152
Capacitance becoming L, 48
Capacitor:
 ceramic, 23
 feedthrough, 25, 48, 60, 218, 264
 ideal, 214
 tantalytic, 49
 "verboten," 45, 245
 water, 240
Cascade matrix, 59
Catastrophies, 113, 148, 276
Ceramic, 21-23
Charge-coupled devices (CCDs), 145
Chicken wire (for shielding), 185
Chopping, 135, 139
Cinder, 83
Circuit breaker, 129
Coax cables, 66, 67, 266
Combination filter/limiter, 268
Compossibility, 40
Compression bonds, 36
Conductivity:
 of soil, 20
 specific, 179
 surface, 85
Conversion:
 MKS into CGS, 184
 of modes, 44
Coordinates, spherical, 27
Contact resistance, 29
Contractors, 229
Coplanning, 142, 143, 148
Control common, 131
Controller, 110
Copper, 34, 53, 85, 184, 192
Corner frequency, 212
Corrosion, 36, 232
Cost:
 vs. benefit, 267
 of bureaucracy, 41
 of no EMC, 107
 of EMP testing, 240
 options, 273

[Cost]
 reduction, 37, 83, 118-124, 266, 272-277
 of retrofitting, 232
Counterpositive measures:
 definition, 2
 examples, 37-55
Coupling (See Transfer)
Crawford cell, 191
Cross-current conduction, 265
Current:
 chopping, 34
 cable leakage, 42, 43, 129
 critical, 101, 102
 eddy-, 213, 216, 270
 fault-, 51
 leakage in filter, 215
 limiting, 165
Customer needs, 251
Cut-off point, 3 dB:
 of cable leakage current, 43
 of filter, 212
 of wave guide, 70
 (see also Corner frequency)
Cut-off, thermal, 147
Cylinder, shielding of, 186

Damping:
 critical, 9
 of eigenresonances, 215
 and Q, 9
 resistive, 220
 (see also Losses)
dB/dt reduction by shields, 84
Decriticalization, 111, 112
Degradation, 23
Delay, 73, 156
Dielectric constant:
 complex, 60, 81
 effective, 21, 22
 fictional, 21, 22
 of MOV material, 263
Differentiation, 153-163
Dimensions, 95
Diodes, current limiting, 165
Dirac pulse, 3
Discharge voltage, 24, 25
Discontinuity:
 in line, 264
 in shield, 190, 191

Edge effect, 23
Effects, counterpositive, 37-55
Eigenresonances (see also resonances), 218
Electricity, static:
 effects, 36
 prevention, 128
 and plastics, 279
Electrolysis, 37
Electrooptics (see also Fiber-optics), 150-152, 155-157, 160, 161
EMC:
 expert, 247
 reputation, 45
EMP, 34, 66, 97, 127, 165, 240
Energy:
 range, 31, 96
 transfer, 61
Engineers:
 cookbook, 118, 120, 237, 248
 digitally trained, 252
 education, 38
 good, 167
Entrainment:
 mechanism, 89
 prevention, 90
 of alpha waves, 163-164
 (see also Synchronization)
Equipotential lines, 22-24
Equivalents:
 of line configurations, 228
 lumped, of lines, 60
Expert witness, 237
Explosion, 96

Failure:
 catastrophic, 88
 mechanism, 25
FATTMESS:
 definition, 32
 differentiation, 153-162
 concerning receptors, 89-99
FCC regulations on digital equipment, 281
Ferrite:
 for baluns, 172
 beads, 50, 60, 61
 cofiring, 281
 cracking, 279

[Ferrite]
 Curie point, 94
 for losses, 213
 in rubber, 270
Fiber-optics, 122, 149, 151
Field:
 concentration, 27, 29
 enhancement, 21, 231
 gradients, 22
 near/far, 252
 nonuniform, 27
 planar, 21
 size effect, 27
Field strength:
 away from transmitter, 77, 258
 electric/magnetic, 76, 77
 ranges, 31
 (see also Field)
Filters for EMC, 197
 ceramic:
  bias effect, 257
  breakdown, 23-27, 281
  patents, 50, 281
  reliability, 26, 281
  structure, 48, 50, 263
  temperature effect, 48, 257
 common mode, 215
 mismatched LC-low-pass
  averaging, 159
  bandpass, 72
  books, 45, 195, 206
  configuration, 207-214
  ferroresonant transformer, 159, 160
  insertion gain, 47, 208, 211-222
  needed losses, 245
  passband behavior, 212
  price of, 212
  RC, 150
  and rise time, 72, 158
  ringing, 7, 47, 215, 243, 280
  slew rate, 158
  theory, 197-206
  transition band, 212
 (see also Lossy lines)
Flammability, 96
Flicker, 103
Flux concentration in spheres, 28
Flux density, 21

Foster's reactance theorem, 60
Frequency:
  beat, 89
  critical:
    for filter, 207
    dependency of elements, 47
  domain, 58-71
  eigenresonances, 76-84
  harmonics, 47
  resonance:
    of line, 253
    sensitivity of receptors, 89, 91
  spectrum, 3, 16
Frost, effect on grounding, 94
F/V converter, 155, 161

Galvanic elements, 93
Gaps, opening under vibration, 37
Gates, 158
Gaussian error function, 74
Glitches, 35
Glow discharge, 21
Gold, pinholes in, 40
Green wire, 134, 253
Grounding:
  of antenna, 270
  and bonding for architects, 232
  configuration, 229
  confusion about, 130
  connection disappears, 37
  by copper plate, 52
  double, 42, 129, 281
  and frost, 94
  and green wire, 134
  ground fault, 51
    interruptor, 238, 254
    voltage, 129, 130
  inter-, 134
  loop problem, 43
  of neutral, 130
  plane missing, 251
  problems, 252, 265, 271
  quasi-, 134
  recommended practice of, 163
  resistance, 29
  signal reference, 132
  single point, 130

[Grounding]
  symbols, 132
  tree, 134
  virtual point, 131
Guarding, 44, 45, 229

Handtools, 253, 254
Hazards:
  in bathtub, 54
  by corrosion,
  critical currents, 101
  by entrainment of alpha waves, 103, 104
  by field concentration, 17, 231
  with explosives, 128
  general, 88
  and green wire, 54
  by leakage current, 53
  prevention:
    in buildings, 232
    by solar cells, 145
    for static electricity, 128
  and safety code, 125-130
  sail boat accident, 50-54
  of school buses, 1
  60 Hz vs. microwave, 91
  in shielded room, 81
  by step voltages, 17, 29, 139, 255
  variety of, 113
  (*see also* Hospital; Step voltage)
HF-sealing as source, 33
Holes:
  effect of, 67, 68
  under ground, 80-84, 277
Hospital, intensive care, 139, 165
Humidity:
  and explosives, 128, 129
  and insulators, 98
  and static electricity, 128

IEC, 242
IEEE:
  continued education course, 252
  standards, 132, 163, 168
Ignition cable, 259

Index

Impedance:
 characteristic, 59, 198, 215, 228
 common, 132
 of free space, 77
 of generators, 199
 interfacial, 198, 199, 218, 219
 of loads, 199
 match of lines, 61
 mismatch of filters, 195-220
 of power supplies, 202
 radial, 78-80
Insertion:
 gain, 7, 46, 206, 213, 217
 loss, calculation of, 204
 ratio, 204
Inductance becoming capacitance, 48
Instability, 262
Interfaces:
 in capacitors, 22
 spherical, 80
 of lines, defined, 205, 206
Interference, 27, 260
Ionization, 27
Isolation, 114, 142, 149, 160
Isolation transformer, 157

Junction rectification, 258

Knauer, 196

Label, misleading, 47-49
Lake, effect, 229
Laminations, 47, 215-216, 265
Lamps:
 fluorescent, 130, 233
 incandescent, 3
Laplace transform, 9-17, 73
Latching of C-MOS, 96
Law of physics, 132
Law suit, 39, 50-54, 237, 238
Leakage, 51, 100, 261
LED, 71
Lightning, 13, 30, 33, 120-128, 135-138, 229, 269
Linearity:
 of electrooptics, 152, 161
 of V/F conversion, 155

Lines, equivalents of, 227
Losses:
 by eddy currents, 213, 216, 270
 in ferrite, 61
 in filters, 213-221
 in lines, 59, 63, 66, 73, 261
 lossy lines, 259, 261, 280
 and Q, 9-12, 98
 and rise time, 12, 72, 74
 (see also Current; Resistance)
Low-pass filters (see Filters)

Machine tool industry, 162
Magnetic fields, 21, 22, 178-185
Mass transportation, 162
Matrix, cascade, 59, 204, 261, 264
Maxwell's equations, 179
Measurement:
 of filters:
  by absorption clamp, 198
  according to CISPR, 198
  insertion loss, 198
  by line stabilization network, 198
  by Mil. Std. 220A, 198, 243-245
  in situ, 196, 253
  worst case, 242, 243, 251
 ratiometric, 161
 shielding problems, 190, 191
 according to VDE, far field, 251
 (see also Testing)
Metals for shielding, 181, 192, 193
Microplasma, 23
Microprocessor, 145, 146
Microwave:
 absorbers, 191, 242
 causing cataracts, 100
Mil. Std. 220A, where acceptable, 245
Mismatch, odd/even, 46, 212
Mode:
 common mode/normal mode, 31, 43, 69, 162, 199-203, 224, 227
 stirring, 191, 242
 polarity, 94
Modulation, 50, 89, 257

MOV, 153, 258
MTBC/MTBF, 121
Musorb, 277

Ni-Fe alloy, 193
Noise generator, 240
Noise guide, 32, 86, 168
Nomographs, 11, 63, 64, 74,
    122, 181, 188, 193, 221
Norway, 130
Notching, 159

Oil in transformer, 25
Op-amp, 158
Options, 153
Order of magnitude (OOM), 12
Oversimplification, 250

Partition:
  of filters, 210-212
  of systems, 149
Permeability, 21, 22, 180, 216,
    225
Phase:
  lag, 158
  measure, 59
  separation, 66
Pigtails, 66
Pitfalls, 37-54
Pipes, 233
Planning, anticipatory, 274
Plastics, 128, 278
Position paper, U.S./Brit., 243
Power, standby, 148
Preamplifier, 145
Precorona, 93
Probability of failure, 123
Problems, 248, 252-276
Propagation measure, 59, 80
Pulse:
  deformation by shield, 84, 85
  destruction by, 97
  discrimination, 158
  errors, 97
  generator, 242
  long, 12, 96
  myographic, 37
  reverse voltage, 93

[Pulse]
  shapes, 12-16, 32, 33
  spectra, 16
  for testing, 81, 240

Q, 10, 32, 196, 219, 242

Raceways, 232
Radiation:
  of lossy ignition cable, 65
  resistance, 76
  by short wire, 267
Rebars, 76, 232
Receptors:
  classification, 87
  living, 100-105
  semiconductors as, 92-99
Recovery:
  time, 35
  voltage, 9
Rectification, 96
Redundancy, 150
Reflection, 80, 264
Reliability, 121
Resistance:
  of human body, 102
  negative, 7, 203
Resistivity:
  of soil, 81
  surface, 85
Resistor:
  constant RF, 61
  ideal, 4
  under pulsed conditions, 43
Resonance:
  of alpha waves, 104
  of C, 49
  of components, 48, 223
  damped, 7-11, 136
  dominant, 9
  eigen, 7, 207, 218, 228
  of feedthrough capacitors, 49,
    261
  impossible, 145
  interfacial, 207, 219, 223
  of L, 48
  and lossy filter, 280
  in MOVs, 263
  parallel/series, 8

Index

[Resonance]
of printed circuit boards, 250
range of, 32, 98
ringing, 7, 215
of shielded rooms, 81, 190
shift in filters, 258
and size, 97
suppression by ferrite, 263
and time constants, 6, 7
Rise time, 71-73, 85
River, 229
Rogowski profile, 22
Rusty bolt, 36

Safety:
codes, 121, 125-130
margin, 245
measures, 133, 134
neutral ground for, 130
wire, 54
(see also Hazards)
Sampling theorem, 3
Saturation, 48, 185, 225
Schools, engineering, 248
SCR bridge affected by ringing, 47
Semiconductors, 34, 37, 93
Self-certification, 282
Sensors, 111, 114, 145, 150
Shielding:
discontinuities, 41, 66, 177-179
earthen, 84, 277
edges in, 190
electric/magnetic fields, 180
guard, 253
holes in, 70
joints, 190
pulses, 85, 191, 192
Shock, 101
Short circuit, 133, 140, 206, 253
Showers, R., 242
Shut-down, 147
Side effects, 102-105
Side flashes, 54, 259
Signals buried in noise, 99
Signal/noise ratio, 99
Silver and sulphur (see also Solder), 190

Sine integral, 73
Site selection, 232
Size:
of ceramic capacitors, 22
definition, as used here, 32
of filters, 212
and resonance, 97
in shielding, 178, 182-184
Skin depth, 67, 83, 180-181, 216
Slew rate, 75, 158, 263
Smith's diagram, 63
Snubber, 171
Software, 145
Soil, 81-84
Solder:
all-around, 48
silver-bearing, 21, 48
Sources:
attributes, 30, 31
categorization, 32
material-caused, 32-33
narrowband, 33
real, 32
Spaces (partition), 111, 144, 149
Spark gap, 172
Spheres:
air in oil, 27
water in oil, 27
as shield, 178
Standards:
and costs, 236
and filters, 242-246
fragmentation, 238
limitative, 215
limits, 38-40, 216
Mil. Std. 220A, 198, 221, 242-245, 251
normative, 235
for specific industries, 163
for static electricity, 279
and thinking, 118
Statistics, 98, 99, 125-127, 146, 198, 235
Steel:
reinforced, 185
Curie point, 94
Step:
function, 3, 4
voltage, 17, 29, 139, 255
Stopband performance, 223

Suck-out point, 49
Suppression:
  localized, 152
  measures, 172-175
Supermalloy, 193
Surfactant, 25
Switching:
  ideal, 3
  delay, 92
Synchronization:
  calculation, 262
  in V/F conversion, 158
  (see also Entrainment)
System:
  center, 131
  classes, 99, 162
  differences, 108-109
  ideal, 150
  large, 132
  speed, 130, 140, 164

Temperature:
  and breakdown, 25, 92
  effect on ceramic feedthrough
    capacitors, 49
  expansion coefficient, 36
  hot spots, 93
  of oil, 25
  and secondary breakdown, 93
  for testing (70°C), 275
  thermoelectricity, 37
Tempest, 195
Test:
  acceptance, 243
  under bias, 243
  for ceramic filters, 66
  for EMP, 81, 240, 241
  impulse, 240
  motor noise, 240
  relay noise, 240
  ringing, 243
  temperature, 25, 240, 275
  under vibration, 98
  worst case, 242-245
  (see also Measurements)
Thinking, 142, 143, 155
Time:
  of breakdown, 91, 92
  constants, 4-9
Time domain consideration for:
  testing in shields, 81, 240, 241
  conducted transfer, 71-75
  radiated transfer, 84-86
Timing, 156-159
Transfer:
  vs. channel, 57
  impedance, 66-68, 273
  phase to phase, 130
  radiated, 75, 76
  spread, 57, 110
Transformer, 8, 157, 253
Transients, 4-7, 34, 141
Transistor, pulse sensitivity, 31
Transition band, 47, 212
Transition layer, 82, 83
Transit time, 140
Transmission line, 58-75
Tranzorbs, 139
Triax cables (see also 271), 281
Triggers, 111
Truth, adversary, 237
Twisted pair, 43

Underwriters, 102, 280

V-curves, 121-123, 272
VDE standards, 238, 253
V/F conversion, 155-161
Vibration, 98
Voltage:
  error, 45
  limiters, 139, 173, 174, 276
  regulator, 160, 229
  reverse, 35
  spike reduction, 280
  step, 17, 229
Voltage standing wave ratio
  (VSWR), 63

Wave guide, 70, 71, 258
Wave front, slowing, 5, 6, 74
Whip antenna, 263
Wire, colored, 129, 135
Wiring large systems, 130

Zener diode, 172
Zoning, 135